OXFORD MEDICAL PUBLICATIONS

INTERFERONS

Their Impact in Biology and Medicine

INTERFERONS
Their Impact in Biology and Medicine

EDITED BY

JOYCE TAYLOR-PAPADIMITRIOU
Imperial Cancer Research Fund

OXFORD NEW YORK TORONTO
OXFORD UNIVERSITY PRESS
1985

Oxford University Press, Walton Street, Oxford OX2 6DP
Oxford New York Toronto
Delhi Bombay Calcutta Madras Karachi
Kuala Lumpur Singapore Hong Kong Tokyo
Nairobi Dar es Salaam Cape Town
Melbourne Auckland
and associated companies in
Beirut Berlin Ibadan Mexico City Nicosia

Oxford is a trade mark of Oxford University Press

Published in the United States
by Oxford University Press, New York

British Library Cataloguing in Publication Data

Interferons: their impact in biology and medicine.—(Oxford medical publications)
1. Interferon
I. Taylor-Papadimitriou, Joyce
574.2'95 QR187.5
ISBN 0–19–261536–X
ISBN 0–19–261481–9 Pbk

Library of Congress Cataloguing in Publication Data

Main entry under title:
Interferons, their impact in biology and medicine.
Includes bibliographies and index.
1. Interferon. I. Taylor-Papadimitriou, Joyce.
[DNLM: 1. Interferons. QW 800 I6236]
QR187.5.I583 1985 574.2'9 84–27353
ISBN 0–19–261536–X
ISBN 0–19–261481–9 (pbk.)

Typeset by Latimer Trend & Company Ltd, Plymouth
Printed in Great Britain by St Edmundsbury Press, Bury St Edmunds, Suffolk

Preface

Twenty-seven years ago, Isaacs and Lindenman described the interferons as a group of proteins and glycoproteins, produced by cells in response to virus infection, which could inhibit the growth of a wide range of viruses. Although early attempts to apply these agents in the clinic were disappointing, investigations continued using *in vitro* culture systems and animal models. During the sixties it became clear that in addition to inhibiting virus growth, interferon preparations could regulate cell growth and function. In particular, they were shown to have an important role as regulators of the immune response, being produced in response to antigenic stimuli.

The interferon preparations which were used in these early studies contained other materials produced or released by virus-infected or antigen-treated cells. Although purification of interferons was obviously important, the work proceeded very slowly, partly because of the difficulty of purifying small amounts of a highly active material from a mixture, and partly through lack of funding. The late Kurt Paucker, who pioneered the purification of mouse interferon, supported his work by selling the product. However, some companies did support small groups to study production and purification of human interferons. Ernest Knight at Dupont and Karl Fantes at Wellcome, spent many man years on this difficult problem. It is probably true to say that, in the early days, most investigators working on the action of interferons produced their own, in small quantities, and used, at best, partially purified material. This situation changed dramatically just a few years ago when the first human interferon genes were cloned. At the same time the virally induced HuIFN-α and HuIFN-β were produced in large quantities and purified. Kari Cantell at the Finnish Red Cross was an important figure in these developments, since he committed himself to developing the large scale production of interferon from virus-induced human leucocytes. It was by using the mRNA from such leucocytes that Weissmann and his colleagues in Zurich were able to isolate the first cDNA clone for a species of human interferon α.

The cloning of the interferon genes and the availability of purified interferons has revolutionized the interferon field. Firstly, a shift in emphasis has occurred so that the molecular biology of the interferon gene families, and the proteins produced by them, has become an important research area. Although much of this work is going on in, or in collaboration with, the biotechnology companies which isolated the cloned genes, some University investigators such as Paula Pitha have produced clones and are using them to study gene expression. A second important consequence of what might be termed the interferon revolution is that purified preparations of single species (or naturally produced mixtured) of human interferons are now widely available. This means that studies following the action of interferons, both in experimental model systems and in patients with disease, can be done by a greater number of scientists and clinicians and the results can be interpreted unequivocally. It turns out that the regulatory functions of the interferons are indeed intrinsic properties of the interferon molecules themselves, making this unique group of proteins of considerable potential interest to investigators working on the functions that they regulate, namely growth and differentiation.

Many of the new recruits to the interferon field are coming from the field of immunology, since it is now known that γ interferons are lymphokines, and that all the interferons are important components of the immune response, presenting the first line of defence against viruses, recruiting and enhancing the function of the effector cells of the immune system and probably modifying the production of other lymphokines. It is of course the effects of the interferons on cell growth and on the function of the effector cells of the immune system which has focused attention on them as potential anti-cancer agents. Indeed, this finding shifted attention away from their antiviral action. Nevertheless important progress has been made in working out mechanisms underlying this action, and now, with the availability of human interferons, clinical trials are under way in the therapy of viral disease as well as in malignancy.

This book attempts to put the recent developments in the interferon field in context, emphasizing those interfaces which are growth areas both in interferon research and in biology and medicine. It is directed not only to interferon specialists but to research scientists and clinicians who may want to initiate investigations. As yet we cannot predict the ultimate role of the interferons in the management and treatment of viral or malignant disease. However, a large number of trials are under way and the results of these trials (some very promising) are presented in the last two chapters of the book. It is clear that whatever the outcome of the clinical trials, the interferons are here to stay. Their use in studies of cell and molecular biology can only expand. Moreover, in learning something of the mechanisms by which interferons inhibit virus and cell growth and regulate cell function, it should be possible to apply them more intelligently in the clinic.

I would like to dedicate this book to the memory of Alick Isaacs, with whom I was privileged to work. Even his unquenchable optimism would not, I think, have envisaged that 'the interferon' would come this far.

London J.T.-P.
June 1984

Contents

Contributors

F. R. Balkwill
Imperial Cancer Research Fund, P.O. Box 123, Lincoln's Inn Fields, London WC2A 3PX, UK.

A. Billiau
The Rega Institute of Medical Research, University of Leuven, B-3000 Leuven, Belgium.

R. Dijkmans
The Rega Institute of Medical Research, University of Leuven, B-3000 Leuven, Belgium.

E. N. Fish
Division of Infectious Diseases, The Hospital for Sick Children and Department of Medical Genetics, University of Toronto, 555 University Avenue, Toronto, Ontario, Canada M5G 1X8.

A. Nethersell
Ludwig Institute for Cancer Research, MRC Centre, The Medical School, Hills Road, Cambridge CB2 2QH, UK.

P. Pitha-Rowe
The Johns Hopkins Oncology Center, 600 North Wolfe Street, Baltimore, MD 21205, USA.

M. Riordan
The Johns Hopkins Oncology Center, 600 North Wolfe Street, Baltimore, MD 21205, USA.

E. Rozengurt
Imperial Cancer Research Fund, P.O. Box 123, Lincoln's Inn Fields, London WC2A 3PX, UK.

G. Scott
Department of Clinical Microbiology, University College Hospital, Gower Street, London WC1E 6DB, UK.

K. Sikora
Ludwig Institute for Cancer Research, MRC Centre, The Medical School, Hills Road, Cambridge CB2 2QH, UK.

J. Taylor-Papadimitriou
Imperial Cancer Research Fund, P.O. Box 123, Lincoln's Inn Fields, London WC2A 3PX, UK.

B. R. G. Williams
Division of Infectious Diseases, The Hospital for Sick Children and Department of Medical Genetics, University of Toronto, 555 University Avenue, Toronto, Ontario, Canada M5G 1X8.

Interferon nomenclature

A standard nomenclature was introduced for the interferons by an International Committee and this is reported in detail in *Nature,* **286,** 110 (1980), and briefly in the first issue of the *Journal of Interferon Research,* 1980. The general classification of the interferons is as follows:

IFN type*	Inducer	Cell type producing the IFN
α	Virus	Leucocytes (probably macrophage component), Lymphoblastoid cell lines
β	Virus	Cells in solid tissues
γ	Antigenic stimulus	T cells (helped by macrophages)

*A prefix such as Hu (human) or Mu (murine) is used to designate the animal species from which the interferon is derived.

Since the cloning of the interferon genes, their nucleotide sequences have been elucidated and it is clear that the original classification of the interferons, made on biological and immunological criteria, fits with the structural data. It is also clear, however, that some types of interferons such as HuIFN-α have several non-allelic variants, which have so far been referred to with subscripts, or capital letters (e.g. HuIFN-α_1 = HuIFNαD; HuIFN-α_2 = HuIFNαA). Moreover, hybrid interferons and interferon analogues have been and are being produced by genetic engineering techniques, and these need to be designated in an organized way. Using the information available on the sequences of the interferon genes, new recommendations have recently been developed for classification and nomenclature of the genes and the proteins they code for. These proposals are outlined in the *Journal of General Virology,* **65,** 669 (1984).

1 An introduction to the genes of the interferon system

R. Dijkmans and A. Billiau

1.1. INTRODUCTION

The molecular cloning of the interferon cDNAs and the isolation of the corresponding genomic clones has brought a wealth of information relevant not only to the interferon structural genes themselves but also to many other facets of the interferon system. The data coming from sequence analysis of these genes, and the proteins they code for, have confirmed the correctness of some of the long-standing assumptions made about interferons and resolved some controversial issues. For example, although it had been clear for a long time that the interferon molecules act from outside the cell, it was the finding that the interferon cDNA codes for a pre-protein with a signal peptide that definitely indicated that interferons are secretory proteins. Similarly, the classification of interferons into IFN-α, IFN-β and IFN-γ, which had been based mainly on serological criteria, received definitive status from the nucleic acid data, which revealed extensive sequence differences between the α, β and γ genes. In fact, protein chemists had been able to resolve within the IFN-α serotype two or more interferon sub-populations with slightly different properties such as species specificity; analysis of cloned interferon genes showed that there do indeed exist different α-interferon genes with different gene products.

In addition to their antiviral effects (Chapters 3 and 6), interferon preparations were found to have multiple effects on cell growth and function (Chapters 4 and 5), but the question of whether these effects were due to the interferon molecules or to other components (produced by cells induced to make interferon) was not clear. This controversy has now been resolved by demonstrating that the same plethora of biological effects can be obtained using purified interferons produced by cells transformed by the cloned interferon genes. The study of the expression of these genes in homologous systems can now be done at the molecular level and such studies throw light not only on the regulation of interferon expression, but on the more general area of gene expression in eukaryotes (see Chapter 2).

Although the cloning of the interferon genes has brought clarity to the field, it has also brought a new degree of complexity. In addition to the several subtypes of HuIFN-α which are naturally produced by leucocytes, hybrid and synthetic genes have been brought to expression. It is clearly a major problem to test all of these interferons (singly and in combination) for differential activity *in vitro,* and it is an even bigger problem to test their usefulness in the clinic. However, the use of the interferon gene analogues and their expression products is providing a growing insight into the conformation and action of interferon proteins.

There are several recent reviews which cover in detail the cloning and sequencing of the interferon genes and their expression in bacteria and yeast (Weissman 1981; Weissman, Nagata, Boll, Fountoulakis, Fujisawa, Fujisawa, Haynes, Henco, Mantei, Ragg, Schein, Schmid, Shaw, Streuli, Taira, Todokoro, and Weidle 1982; Collins 1983, Kingsman and Kingsman 1983). Here we will briefly discuss the interferon structural genes and proteins in the context of the interferon system and introduce the other genes and gene products which appear to be relevant to the expression and action of the interferons.

1.2. BASIC STRATEGIES USED IN THE ISOLATION, CHARACTERIZATION, AND MANIPULATION OF THE STRUCTURAL INTERFERON GENES

Between 1979 and 1982, DNA sequences carrying the complete genetic code for human interferons α, β and γ were obtained by molecular cloning techniques. In each case the same general strategy was followed, largely based on the pioneering work of Taniguchi and colleagues who obtained the first clone for an interferon gene (HuIFN-β; Taniguchi, Sakai, Fujii-Kuriyama, Muramatsu, Kobayashi, and Sudo 1979), and it is appropriate to briefly outline the methods used.

1.2.1. Cloning of interferon cDNAs

In order to begin to isolate interferon mRNA it was first necessary to establish the kinetics of induction of this material. Suitable human cells were appropriately induced (see Chapter 2) to produce one or more of the interferon proteins. The presence of interferon-specific mRNA in the cells was demonstrated and the kinetics of its formation followed, usually by injecting RNA extracted from the cells into *Xenopus* oocytes and showing the appearance of biologically active interferon in the incubation medium. In other instances cell-free translation systems (rabbit reticulocyte or wheat-germ lysates) were used; in this case translation was monitored by electrophoresis of the translation products showing formation of a labelled protein band having the anticipated molecular weight (~20 k) and being immunoprecipitable with the appropriate antibody.

Once the kinetics of mRNA formation were established, a large batch of such RNA could be prepared and purified by chromatography on oligo-dT affinity columns, which removed non-polyadenylated RNAs, and by centrifugal fractionation, allowing the isolation of fractions that were rich in interferon-specific mRNA; only a minor percentage of the mRNA that was isolated was however interferon-specific.

Using the now-classic reactions, the RNA preparations enriched for IFN mRNA were transcribed *in vitro* into linear double-stranded DNA sequences of various lengths and specificity. When part of the amino acid sequence of an interferon became known, the corresponding oligonucleotides could be synthesized and used as primers of the reverse transcriptase, thus exerting selective pressure in favour of interferon sequences. Indeed based on the published sequence of the HuIFN-α_1 gene, Edge and his colleagues were able to chemically synthesize the complete gene from oligonucleotides (Edge, Greene, Heathcliffe, Meacock, Schuch, Scanlon, Atkinson, Newton, and Markham 1981).

The cDNA preparations obtained by reverse transcription of preparations of mRNA consisted of populations of molecules only a minority of which were interferon-specific. The termini of these molecules were enzymatically tailored to make them suitable for insertion into a circular plasmid DNA that had been opened and trimmed to allow base pairing of the ends with those of the cDNA molecules. Classical bacterial cloning using antibiotic-resistant markers on the plasmid allowed the isolation of bacterial clones carrying plasmid cDNA chimeras. Since the great majority of these clones contained DNA sequences not related to interferon, techniques were required for the selection of those few clones carrying the interferon cDNA. The most simple technique is hybridization with radioactively labelled DNA that already contains interferon-specific sequences. Such DNA may be available as oligonucleotides synthesized from a known amino-acid sequence or as cloned DNA from a related interferon. In the beginning such DNAs were not available and selection had to be done by testing hybridization to mRNA preparations containing the interferon message. In a hybridization – translation assay DNA from individual or groups of clones is immobilized on filter paper and hybridized with IFN mRNA-containing samples. The mRNA which specifically hybridized was then eluted and tested for its ability to code for IFN in the oocyte injection assay. This method was used by Taniguchi and colleagues to identify the IFN-β clones amongst those colonies selected by a colony hybridization method (Taniguchi, Sakai, Fujii-Kuriyama, Muramatsu, Kobayashi, and Sudo 1979). Gray and colleagues used the difference in hybridization with cDNA derived from induced cultures and with cDNA from non-induced cultures as a direct way to select the correct clones coding for HuIFN-γ (Gray, Leung, Pennica, Yelverton, Najarian, Simonson, Derynck, Sherwood, Wallace, Berger, Levinson and Goeddel, 1982). In the case of the HuIFN-α genes, the selection for positive clones was made easier by the discovery that some of the *E. coli* clones were producing low levels of biologically active interferon and screening could be done by assaying for antiviral activity (Weissman 1981). Sequence analysis of the cDNA clones has given the nucleotide sequence of the coding as well as the noncoding regions of the corresponding mRNA, and from this the amino acid sequences of the native interferon proteins could be derived.

1.2.2. Isolation and analysis of genomic DNA clones

By using cloned cDNAs as probes, the genomic organization of interferon-coding sequences has been analysed, and the structure of the interferon genomic DNA has been elucidated. Moreover, genomic sequences were identified in bacteriophage and/or cosmid clone bands of human DNA. Analysis of these genomic clones has shown that for certain serotypes of interferon the genome contains not one gene, but rather a gene family. (Nagata, Brack, Henco, Schambock, and Weissmann 1981, Weissman *et al.* 1982; Collins 1983). Also, the presence, number, size and location of introns could thus be determined. Finally, the 3′- and 5′- flanking regions of the gene could be sequenced, a prerequisite for identifying the signals acting as regulatory elements for transcription and induction (Chapter 2).

1.2.3. Construction of artificially modified interferon genes

Invariably the interferon genes have been expressed in *E. coli* following their insertion in suitable *E. coli* expression plasmids. Some of the interferon genes were also inserted into vectors which allowed their expression in heterologous mammalian cells or in yeasts (Gray *et al.* 1982; Kingsman and Kingsman 1983). Aside from the manipulations necessary to meet the requirements for expression, various other artificial modifications of the interferon genes have been performed. Thus it has been possible to construct hybrid genes consisting of the amino terminal part of one interferon linked to the carboxyl terminal part of another one, such that a full-size interferon results. Another approach has been to introduce targeted mutations aimed at the planned substitution of one particular amino acid in the expressed product by another. This is relatively easy to do when synthesizing the gene from oligonucleotides which may be modified at will. Such replacements have proved useful in generating interferons with altered physicochemical properties.

1.3. STRUCTURE AND VARIABILITY OF THE IFN-GENES

Interferon genes of several animal species (man, mouse, rat, cow) have been cloned and analysed. Of all these, the human interferon genes are at present the best known and some characteristics of these genes and the proteins they code for are listed in Table 1.1.

1.3.1. Human interferon-α

There are at least 13 human IFN-α genes and six pseudogenes with a nucleotide sequence homology of ±90 per cent. They are clustered on chromosome 9 and are devoid of introns. The gene product is a polypeptide

Table 1.1. *General properties of the human interferon genes*

Interferon serotype	Number of genes	Introns	Length of mature protein	Length of signal peptide	Glycosylated	Intramolecular cystein bridges
α[a]	at least 13	0	165, 166	23	no	yes
β	1[b]	0	166	21	yes	yes
γ	1	3	146	20	yes	no[c]

[a] Subtypes have been referred to as $\alpha_1 \alpha_2$ etc. by Nagata, Mantei, and Weissmann (1980) and αA, αB, αC etc. by Goeddel, Leung, Dull, Gross, Lawn, McCandliss, Seeburg, Ullrich, Yelverton, and Gray (1981). α_1 and α_D differ by only one amino acid as do α_2 and α_A. IFN B, C, G, and H are like αD and have also been designated α_1. For a detailed comparison of the sequences see Weissman 1981. IFN-αA and IFN-αD are the predominant interferons in interferon preparations made from leucocytes from the KG-1 myeloblast cell line and from Namalwa cells induced with Sendai Virus (Wellferon).

[b] Up to this date only one HuIFN-β gene has been recognized. However the existence of a second gene has been inferred from studies on mRNA populations (Sehgal and Sagar 1980; Weissenbach, Chernajovsky, Zeevi, Schulman, Soreg, Nir, Wallach, Perricaudet, Tiollas, and Revel 1980).

[c] The only two cystein residues in HuIFN-γ are located on position 1 and 3. Although there are no other proteins known which have a disulphide bond between cysteins separated by only one amino acid, in the case of HuIFN-γ such a bond would not be impossible (M. De Ley, personal communication).

of 188 and 189 amino acids, the first 23 of which constitute the hydrophobic signal peptide. This peptide is split off during transport out of the cell, yielding the active mature form of interferon-α which consists of 165 or 166 amino acids. Some of the naturally produced interferons are missing the last 10 amino acids (at the carboxyl end), which do not appear to be necessary for biological activity (Wetzel, Levine, Estell, Shire, Finer-Moore, Stroud, and Bewley 1982; Arnheiter, Ohno, Smith, Gutte, and Zoon 1983). Cysteine residues are found in the IFN-α polypeptide: both cys29 and cys138 and cys1 and cys90 are linked by disulphide bridges. No possible *N*-glycosylation sites (Asn-X-Thr; Asn-X-Ser) are present; however *O*-glycosylation can not be excluded.

The interferon-α subtypes have a different target cell specificity and can be assembled in two or three subfamilies. (Weissman *et al.* 1982). In natural interferon preparations the subtypes arise in varying ratios, (Allen and Fantes 1980; Rubenstein, Rubinstein, Fomillette, Miller, Waldman, and Pestka 1979). The 5′ noncoding region of the IFN-α mRNA is ±70 nucleotides long. The 3′ noncoding region is variable in size and contains the polyadenylation signal (AAUAAA or a related sequence).

1.3.2. Human interferon-β

Until now only one human IFN-β gene has been recognized; like the IFN-α genes it lacks introns and is located on chromosome 9. Although the HuIFN-α and HuIFN-β amino acid sequences show only 29 per cent homology, the general structure of the IFN-β gene and of the 5′ and 3′ flanking regions has much in common with those of IFN-α. The gene codes for a protein of 187 amino acids with a signal peptide of 21 amino acids. However, IFN-β has a potential *N*-glycosylation site and is more hydrophobic than IFN-α. Moreover, the polypeptide contains only three cysteine residues of which cys31 and cys141 are linked in a disulphide bridge.

1.3.3. Human interferon-γ

There is little resemblance between the unique IFN-γ gene and the IFN-α and -β genes. The IFN-γ gene is located on chromosome 12 and has three introns in the coding region. The 5′ untranslated sequence is ±110 nucleotides long, the 3′ untranslated region is 587 nucleotides long and contains the AAUAAA polyadenylation signal. The gene codes for a protein of 166 amino acids of which 20 constitute the putative signal pepthide. The mature protein is quite basic in character and is probably glycosylated. The presence of inter- and intramolecular disulphide bridges cannot be ruled out.

1.4. STRUCTURE OF AND RELATIONSHIPS BETWEEN INTERFERON PROTEINS AS DERIVED FROM THEIR NUCLEIC ACID SEQUENCES

The structural information coming from the molecular cloning of the interferon genes can help in understanding how interferon interacts with cells. Because IFN-γ uses a cell receptor different from that used by IFN-α and -β, one may postulate that they will differ in regions that bind the receptor. On the other hand, because different interferons have similar cellular reaction mechanisms they may have been derived by divergence from common ancestral molecules and this should be reflected in conservation of certain regions in the molecule. Alternatively, different interferons may be derived from different ancestors and may have acquired similar features by convergence, which may also be reflected in similarities in amino acid sequence. To find such similarities or divergences, one can simply compare the amino acid sequences of different interferons from the same species (Taniguchi, Mantei, Schwarzstein, Nagata, Muramatsu, and Weissmann 1980). The molecular cloning of interferons from different species, however, has helped detect structurally important regions of the interferon molecules. The cDNA sequences of mouse IFN-α (MuIFN-α), rat IFN-α, mouse IFN-β (MuIFN-β) and mouse IFN-γ (MuIFN-γ) are available (Shaw, Boll, Taira, Mantei, Lengyel, and Weissman 1983; Dijkema, Pouwels, de Reus, A., and Schellekens 1984; Higashi, Sokawa, Watanabe, Kawade, Ohno, Takoaka, and Tanaguchi 1983; Gray and Goedell 1983). The general features of these genes are basically similar to these of the human interferon genes. Mouse and rat IFN-α are also coded for by a family of genes (Wilson, Jeffreys, Barrie, Boseley, Slocombe, Easton, and Burke 1983). No introns could be detected and the gene products show a high degree of homology at the amino acid level. However, in rat and mouse IFN-α potential *N*-glycosylation sites can be recognized.

MuIFN-β is coded by a single gene and has three potential *N*-glycosylation sites and a single cysteine residue. MuIFN-γ has the same genomic organization as HuIFN-γ (e.g. three introns). Both polypeptides have an excess of basic residues and contain two potential *N*-glycosylation sites. However, MuIFN-γ has three rather than two cysteine residues (positions 1, 3, and 136) and is 10 amino acids shorter. An overview of homology data of mouse and human interferons is given in Fig. 1.1. It is striking that interferon-γ, which has the most strict species specificity, also has the lowest interspecies homology, whereas interferon-α has considerable activity on heterologous cells and higher inter-species homology. In general, interferons have a greater interspecies than inter-type homology.

HuIFN-α and HuIFN-β show homology at 29 per cent of the amino acid positions. The conservation is highest in two domains of the mature molecule: between amino acids 28 and 40, and between amino acids 120 and 150. The same highly conserved domains are revealed by comparing the amino acid sequences of the IFN-α subtypes. The amino acid sequences of HuIFN-α and MuIFN-α share 60 per cent of their positions. However, if the amino acids between positions 28 and 40 and between 122 and 150 are compared, homologies of 90 per cent and 80 per cent respectively are found. Approximately the same percentages are found when rat IFN-α is compared with HuIFN-α (Fig. 1.2). The conservation of these domains suggests that

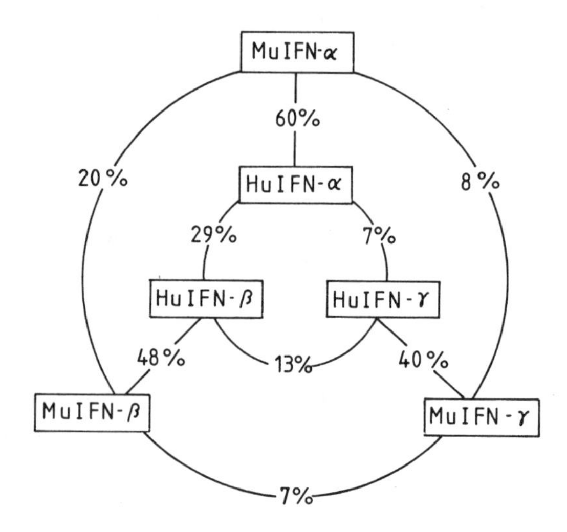

Fig. 1.1. Inter-type and inter-species homologies between interferons.

they are important for the activity of both interferon-α and interferon-β.

An interesting characteristic of the second highly conserved domain is that it shows some homology with the β subunit of cholera toxin (see Fig. 1.2). Cholera toxin is able to interfere competitively with the binding of interferon on the cell membrane. This could be an indication that the region between amino acid 120 and 150 is involved in the binding of interferon to the receptor of the target cell.

On superficial examination, IFN-γ seems to be completely different in structure from IFN-α and IFN-β. However, one domain of the IFN-γ polypeptide that should have a high potential to form an amphiphilic helix (59–104) has ±30 per cent homology with IFN-β (66–112) (De Grado, Wasserman, and Chowdhry 1982). A similar homology is found between MuIFN-β and MuIFN-γ (Fig. 1.3). If gaps are introduced in the sequence small homologous regions (two or three amino acids) can also be found elsewhere in the molecule. The significance of this homology is not completely obvious. It is striking that the region of highest homology between IFN-β and IFN-γ is different from the most conserved domains in IFN-β and IFN-α.

By comparing the nucleotide sequences it is possible to gain some insight into the evolution of the interferon genes. If it is assumed that single-copy DNA mutates at a rate of 0.1% per million years, then the IFN-α gene family should have arisen about 50 million years ago. A progenitor of the interferon-α gene family and a precursor of the modern IFN-β gene should have diverged from a common ancestral gene about 500 million years ago (Taniguchi, Mantei, Schwarzstein, Nagata, Muramatsu, and Weisman 1980; Gillespie, Pequignot, and Carter 1984). Whether IFN-γ diverged still earlier in evolution or had its own origin is not yet known. In any case it seems

Sequence	Start													
HuIFN-α_1	28	*Ser*	*Cys*	*Leu*	Lys	*Asp*	*Arg*	*His*	*Asp*	*Phe*	Gly	Phe	*Pro*	*Glu*
MuIFN-α_1	28	*Ser*	*Cys*	*Leu*	*Lys*	*Asp*	*Arg*	Lys	*Asp*	*Phe*	*Gly*	*Phe*	*Pro*	Glu
Rat IFN-α_1	28	Ser	Cys	Leu	Lys	Asp	Arg	Lys	Thr	Phe	Gly	Phe	Pro	Leu
HuIFN-β_1	30	Tyr	Cys	Leu	Lys	Asp	Arg	Met	Asn	Phe	Asp	Ile	*Pro*	Glu
MuIFN-β_1	28	Ile	Asn	Leu	Thr	Tyr	Arg	Ala	Asp	Phe	Lys	Ile	Pro	Met

Sequence	Start																
HuIFN-α_1	120	*Val*	Lys	*Lys*	*Tyr*	Phe	*Arg*	*Arg*	*Ile*	*Thr*	*Leu*	*Tyr*	*Leu*	Thr	*Glu*	Lys	*Lys*
MuIFN-α_1	120	*Val*	*Arg*	*Lys*	*Tyr*	*Phe*	*His*	*Arg*	*Ile*	*Thr*	*Val*	*Tyr*	*Leu*	*Arg*	*Glu*	*Lys*	*Lys*
Rat IFN-α_1	120	Val	Arg	Glu	Tyr	Phe	His	Arg	Ile	Thr	Val	Tyr	Leu	Arg	Glu	Asn	Lys
HuIFN-β_1	122	Leu	Lys	Arg	Tyr	Tyr	Gly	Arg	Ile	Leu	His	Tyr	Leu	Lys	Ala	Lys	Glu
MuIFN-β_1	117	Leu	Lys	Ser	Tyr	Tyr	Tyr	Arg	Val	Gln	Arg	Tyr	Leu	Lys	Leu	Met	Lys
Cholera toxin	67	Arg	Met	Lys	Asn	Thr	Leu	Arg	Ile	Ala	*	Tyr	Leu	Thr	Glu	Ala	Lys

Sequence	Start																
HuIFN-α_1	136	*Tyr*	*Ser*	Pro	*Cys*	*Ala*	*Trp*	*Glu*	*Val*	*Val*	*Arg*	*Ala*	*Glu*	*Ile*	*Met*	*Arg*	*Ser*
MuIFN-α_1	136	*His*	*Ser*	*Pro*	*Cys*	*Ala*	*Trp*	*Glu*	*Val*	*Val*	*Arg*	*Ala*	*Glu*	*Val*	*Trp*	*Arg*	*Ala*
Rat IFN-α_1	136	His	Ser	Pro	Cys	Ala	Trp	Glu	Val	Val	Lys	Ala	Glu	Val	Trp	Arg	Ala
HuIFN-β_1	138	Tyr	Ser	His	Cys	Ala	Trp	Thr	Ile	Val	Arg	Val	Glu	Ile	Leu	Arg	Asn
MuIFN-β_1	133	Tyr	Asn	Ser	Tyr	Ala	Trp	Met	Val	Val	Arg	Ala	Glu	Ile	Phe	Arg	Asn
Cholera toxin	82	Val	Glu	Lys	Leu	Cys	Val	Trp	Asn	Asn	Lys						

Fig. 1.2. Sequence homology of two conserved domains in IFN-α and IFN-β. The closed boxes indicate positions which are common in all published IFN-α and IFN-β sequences. IFN-α sequences that are homologous in all subtypes (excluding the pseudogene HuIFN-αE) are underlined. The sequence of the β subunit of cholera-toxin that shows some homology with HuIFN-α/β is also given.

HuIFN-β	66Ile	Phe	Ala	Ile	Phe	Arg	Gln	Asp	Ser	Ser	Ser	Thr	Gly	Trp	Asn	Glu	Thr	Ile	Val	Glu	Asn	Leu	Leu	Ala	Asn
HuIFN-γ	59Leu	Phe	Lys	Asn	Phe	Lys	Asp	Asp	Gln	Ser	Ile	Gln	Lys	Ser	Val	Glu	Thr	Ile	Lys	Glu	Asp	*	Met	Asn	Val
MuIFN-β	62Val	Phe	Leu	Val	Phe	Arg	Asn	Asn	Phe	Ser	Ser	Thr	Gly	Trp	Asn	Glu	Thr	Ile	Val	Val	Arg	Leu	Leu	Asp	Glu
MuIFN-γ	59Leu	Phe	Glu	Val	Leu	Lys	Asp	Asn	Gln	Ala	Ile	Ser	Asn	Asn	Ile	Ser	Val	Ile	Glu	Ser	His	*	Leu	Ile	Thr

HuIFN-β	91Val	Tyr	His	Gln	Ile	Asn	His	Leu	Lys	Thr	Val	Leu	Glu	Glu	Lys	Leu	Glu	Lys	Glu	Asp	Phe	Thr
HuIFN-γ	83Lys	Phe	Phe	Asn	Ser	Asn	Lys	Lys	Lys	Arg	Asp	Asp	Phe	Glu	Lys	Leu	Thr	Asn	Tyr	Ser	Val	Thr
MuIFN-β	87Leu	His	Gln	Gln	Thr	Val	Phe	Leu	Lys	Thr	Val	Leu	Glu	Glu	Lys	Gln	Glu	*	Glu	Arg	Leu	Thr
MuIFN-γ	83Thr	Phe	Phe	Ser	Asn	Ser	Lys	Ala	Lys	Lys	Asp	Ala	Phe	Met	Ser	Ile	Ala	Lys	Phe	Glu	Val	Asn

Fig. 1.3. Region of high homology between IFN-β and IFN-γ (partially derived from De Grado, Wasserman, and Chowdhry 1982).

probable that the different interferon serotypes did exist before the mammalian radiation (about 70 million years ago). Therefore, it can be expected that all mammalian species have all three interferon types.

Another stretegy used to delinate the important regions of the interferon molecule is the construction and expression of hybrid cDNA molecules. Thus, a hybrid IFN molecule was prepared consisting of the amino terminal third of IFN-α_1 and the carboxy terminal two-thirds of IFN-α_2 and vice versa (Streuli, Hall, Stewart, Nagata, and Weissmann 1981). Both parts of the molecule seemed to play a role in the binding on the cell surface receptor, but binding to homologous and heterologous cells appeared to depend on different parts of the molecule. Alternatively, a synthetic interferon gene can be designed with a limited number of amino acid substitutions. By comparing the biological activity of the expression product with the activity of the natural interferon molecule it can then be concluded what part of the molecule is responsible for a particular biological effect. For instance in HuIFN-γ the unique tryptophane residue and the methionine residue at position 48 seem to be indispensable for the antiviral activity (Alton, Stabinsky, Richards, Ferguson, Goldstein, Attrock, Miller, and Stebbing 1983). Attempts to prepare biologically active peptides either by synthesis or selective enzymic degration of α-IFNs suggested that most of the molecule was involved in the biological activity (Wetzel *et al.* 1983).

A few features of the interferon genes and their coded proteins need more attention.

1.4.1. Disulphide bonds

There are four or five cysteines in HuIFN-α, six in MuIFN-α, three in HuIFN-β, one in MuIFN-β, two in HuIFN-γ, and three in MuIFN-γ. Disulphide bridges play a principal role in the structure of proteins, and mature HuIFN-α treated with reducing agents loses its antiviral activity, but HuIFN-γ does not. Disulphide bridges can also contribute in the dimerization of polypeptides and may be involved in forming stable oligomers of interferon preparations. Reversible aggregation to form dimers and higher oligomers may also occur without S–S bond formation (Shire 1983) and it has been suggested that HuIFN-γ exists as a tetramer, HuIFN-β as a dimer and HuIFN-α as a monomer (in dilute solution) (Pestka, Kelder, Familletti, Moschera, Crowl, and Kempner 1983); the nature of the intramolecular bond in the oligomers is not clear. The bridge between the cysteines in positions 29 and 138 was thought to be crucial to the biological activity of HuIFN-α since selective breakage of this link destroys activity (Wetzel *et al.* 1983). However, recent data showing that a truncated peptide consisting of the amino terminal 110 amino acids has antiviral activity argues against this. (Ackerman, Nedden, Heintzelman, Hunkapiller, and Zoon 1984).

1.4.2. Glycosylation

Glycosylation is probably not essential for the biological activities of interferons. Thus human IFN-α is not *N*-glycosylated but the mouse and rat IFN-αs are. Also interferons produced in bacteria do not bear carbohydrate moieties but are still biologically active. Finally when natural interferons are treated to remove their carbohydrate groups, they retain their antiviral activity. Possibly glycosylation plays a role in the tissue specificity of interferons, and may affect their stability in plasma.

1.4.3. Introns

Interferon-α and -β genes are devoid of introns. This feature is quite unusual in eukaryotic proteins. One hypothesis to explain this is that the splicing of genes is impaired or precluded by viral infection. If so it would constitute an evolutionary advantage for cells not to need a splicing process for interferon synthesis. The interferon-γ gene, on the other hand, does possess introns, possibly because it is induced by the mitogenic stimulus and antiviral activity is probably not its principal role.

1.5. OTHER GENES OPERATIVE IN THE INTERFERON SYSTEM

Interferons and their genes are only one element of a large system of genes and gene products. This interferon system comprises two main compartments: induction and action. *Induction* of interferon implies activation not only of the interferon genes themselves but also of several other genes, some of which give 'by-products' of interferon preparations, while others yield proteins which regulate transcription and translation of the interferon genes. *Action* of interferon on cells requires the presence of a membrane receptor complex which is coded for by a cellular gene (or genes). Interaction of interferon with this receptor complex results in activation of several genes directing the synthesis of proteins, some of which mediate the biological effects of interferon.

A complete understanding of the interferon system will ultimately require the molecular cloning of all these associated genes and such work is in progress in a number of laboratories. Meanwhile, only fragmentary information is available, mainly from studies with mouse/human cell hybrids and from Mendelian analysis of mouse strains. In somatic cell hybridization cells from different species are fused. The resulting hybrid cell contains chromosomes from both the parental cells but usually some of the chromosomes are lost. It is possible to generate cell lines that have all the mouse chromosomes but lack all but one human chromosome. If one only focuses on the human proteins of the cell, the genetic complexity of the original human cell is reduced 24 times.

Some mouse strains lack a specific interferon-related property or possess it to an extraordinary extent: in these cases an insight into the interferon-associated genes and their functions can be obtained by genetic comparison of these strains with their 'normal' counterparts.

1.5.1. The If loci in mice

The level of interferon produced in response to a specific inducer is genetically determined as is illustrated by the fact that Newcastle Disease virus (NDV) and Sendai virus induce 10-fold higher levels of serum IFN in C57B1/6 than in BALB/c mice (De Maeyer and De Maeyer-Guignard 1979). Moreover, Mendelian analysis indicates that interferon induction by NDV is regulated by another gene locus than when Sendai virus is the inducing agent. These loci are respectively called If-1 and If-2. The exact role of the gene products of these loci in the interferon system is not clear. In fact, the case for designating these genes with the initials of interferon is rather weak, since they may code for proteins that are NDV or Sendai virus-specific rather than interferon-specific. In the meantime, additional similar loci have also been described.

1.5.2. Genes for proteins co-induced with IFN-*β*

Treatment of cells with double-stranded RNA in the presence of metabolic inhibitors results not only in synthesis of interferon but also of several other proteins. This was clearly indicated by the analysis of mRNAs isolated from induced cells that were absent in non-induced cells (Raj and Pitha 1980). Some of the *in vitro* translation products of these RNAs resembled IFN-β in their reaction with certain anti-IFN-β antisera; since they also triggered the antiviral state in appropriate cell systems, and induced 2-5A synthetase activity they were considered to be IFN-β proteins and designated IFN-β_2 etc. (Sehgal and Sagar 1980; Weissenbach, Chernajovsky, Zeevi, Shulman, Soreg, Nir, Wallach, Perricaudet, Tiollais, and Revel 1980). The nucleic acid sequence analysis of cDNA clones corresponding to the co-induced protein, described by Weissenbach, showed a gene structure that was completely different from that of IFN-α and IFN-β although a certain degree of homology was observed in the region between amino acids 45 and 52 of IFN-α/β. So far no cDNA clones corresponding to the proteins described by Sehgal have been reported. Other proteins co-induced with interferon-β have no interferon-like activities (Content, DeWit, Pierard, Derynck, De Clercq, and Fiers 1982; Hauser, Gross, Bruns, Hockkeppel, Mayr, and Collins 1982), but some of them are coded for by genes closely linked to the chromosomal gene of IFN-β.

Studies on the induction mechanism of IFN-β in human or rabbit cells have suggested the existence of a gene which is co-activated with the IFN-β gene and whose product is a repressor protein which inhibits further transcription of the interferon gene. (For review see Burke 1983).

1.5.3. Genes for proteins co-induced with IFN-γ

IFN-γ is induced in lymphocytes by stimulation with mitogens or antigens. Together with IFN-γ these lymphocytes produce various amounts of other lymphokines (e.g. interleukin-2). The genes of several of those proteins have been cloned and analysed. A special case is the 22 k protein described by Van Damme, Billiau, De Ley, and De Somer (1983), which was isolated as a protein co-induced with IFN-γ in concanavalin A-stimulated human leucocytes. This protein had antiviral activity on human fibroblasts and was inactivated by anti-IFN-β antisera, thereby suggesting that it was similar to HuIFN-β. However, its different host range, its different behaviour on affinity columns and the lack of inactivation of IFN-β by anti-22k-antiserum clearly distinguished it from classical HuIFN-β and suggested that 22k could rather be an inducer of IFN-β. This hypothesis was supported by hybridization studies with IFN-β cDNA on mRNA isolated from 22k-induced fibroblasts.

1.5.4. Gene(s) for the receptor of interferon

It is now well established that interaction with membrane receptors is the crucial initial step in interferons induction of an antiviral state or in generating signals which result in effects in growth and function. (Aguet 1980). Analysis of mouse/human cell hybrids suggested that at least one gene coding for the human IFN-α/β cell receptor is located on the long arm of chromosome 21. It was also reported that fibroblast cells, trisomic for chromosome 21, were more sensitive than normal cells to the antiviral effect of IFN-α and IFN-β.

Evidence that the gene on chromosome 21 codes directly for the receptor or regulates its synthesis came from the observation that trisomic 21 cells could bind more radioactively labelled interferon than normal cells (Epstein and Epstein 1982). Although the gene(s) for the interferon receptor(s) is (are) not yet cloned, this should be feasible, if difficult because of the low level of receptor expression.

1.5.5. Genes for interferon-induced proteins

The appearance of certain enzymatic activities in cells treated with interferon indicated that new proteins are expressed or that inactive polypeptides are transformed to active enzymes. Of these enzymes a ppp(A2′p)$_n$A synthetase (shortly 2-5A synthetase) and a protein kinase are probably the best known. These inducible enzymes are almost certainly involved in the antiviral effect of interferons, at least with some viruses (Chapter 3); it is less clear whether they are involved in the effects on cell growth and function (Chapter 5). The mRNA for 2-5A synthetase has been isolated and cDNA clones containing part of the 2-5A synthetase coding sequence have been obtained (Chebath 1983).

Another way to reveal the presence of induced proteins is the comparison by two-dimensional gel electrophoresis of proteins isolated from interferon-treated and mock-treated cells. These studies indicated that IFN-α induces at least 10 new proteins while IFN-γ induces the synthesis of six additional polypeptides, which are not affected by IFN-α (Weil, Epstein, Epstein, and

Grissberg 1983). cDNA clones have been obtained for some of these interferon-induced proteins (Chebath 1983; Samantha 1983).

In the mouse system a gene called Mx has been described (Horisberger, Staeheli, and Haller 1983) that is responsible for an interferon-induced antiviral state selective for orthomyxoviruses. Cells carrying the Mx gene can be protected against influenza virus (but not other viruses) by much smaller doses of IFN than are necessary to protect cells without Mx. The product of the Mx gene is probably a 72k protein. Studies are in progress to clone the cDNA of the Mx gene and this will probably be achieved with the help of cDNAs from mice homozygous for Mx and mice that lack the gene.

1.6. CONCLUDING REMARKS

Ten years ago, the cloning of the structural genes coding for the interferons seemed very far away. One reason for the accelerated development is probably related to the fact that commercial financial commitment acted synergistically with academic excellence in the newly created biotechnology companies. The result is that now, studies on the structural human genes are well and truly launched and the field is expanding to include interferons of other mammalian species and genetically engineered analogues of the human genes. At the time of writing this book, studies on the other genes related to the interferon system (namely those coding for proteins relevant to interferon action or co-induced with the interferons) although not so well advanced, are well established. If these studies progress with anything like the rate of development seen in the field in the last five years then it may be necessary in 1988 to write, not an introductory chapter, but a book on the genes of the interferon system.

Acknowledgements

Studies on interferon in the authors' laboratory are supported by the Cancer Research Foundation of the Belgian A.S.L.K. (General Savings and Retirement Fund) and by the 'Geconcerteerde Onderzoeksacties'. Roger Dijkmans is fellow of the Belgian I.W.O.N.L./I.R.S.I.A. The authors thank C. Callebaut and O. Van Brusselen for editorial help.

1.7. REFERENCES

Ackerman, S. K., Nedden, D. Z., Heintzelman, M., Hunkapiller, M., and Zoon, K. (1984). Biologic activity in a fragment of recombinant human interferon-α. *Proc. Nat. Acad. Sci. USA* **81,** 1045–7.

Aguet, M. (1980). High-affinity binding of ^{125}I-labelled mouse interferon to a specific cell surface receptor. *Nature* **284,** 459–61.

Allen, G. and Fantes, K. H. (1980). A family of structural genes for human (lymphoblastoid leukocyte-type) interferon. *Nature* **287,** 408–11.

Alton, K., Stabinsky, Y., Richards, R. Ferguson, B. Goldstein, L., Altrock, B., Miller, L., and Stebbing, N. (1983). Production, characterisation and biological effects of recombinant DNA derived from IFN-α and IFN-γ analogs. In *The biology of the interferon system 1983,* (ed. E. De Maeyer and H. Schellekens) pp. 119–28. Elsevier, Amsterdam.

Arnheiter, H., Ohno, M., Smith, M., Gutte, B., and Zoon, K. C. (1983). Orientation of a human leukocyte interferon molecule on its cell surface receptor: carboxyl terminus remains accessible to a monoclonal antibody made against a synthetic interferon fragment. *Proc. Nat. Acad. Sci. USA* **80,** 2539–43.

Burke, D. C. (1983) The control of interferon formation. In *Interferons, from molecular biology to clinical application. SGM Symposium 35* (ed. D. C. Burke and A. G. Morris) pp. 67–88. Cambridge University Press, Cambridge.

Chebath, S., Merlin, G., Benech, P., Melz, R., and Revel, M. (1983). Characterisation of cDNA clones for the human (2'–5') oligo A synthetase and for a 56 000 Mr protein induced by interferon. In *The biology of the interferon system 1983* (ed. E. De Maeyer and H. Schellekens) pp. 223–9. Elsevier, Amsterdam.

Collins, J. (1983). Structure and expression of the human interferon genes. In *Interferons, from molecular biology to clinical application. SGM Symposium 35* (ed. D. C. Burke and A. G. Morris) pp. 35–65. Cambridge University Press, Cambridge.

Content, J., De Wit, L., Pierard, D., Derynck, R., De Clercq, E., and Fiers, W. (1982). Secretory proteins induced in human fibroblasts under conditions used for the production of interferon-β. *Proc. Nat. Acad. Sci. USA* **79,** 2768–72.

De Grado, W. F., Wasserman, Z. R., and Chowdhry, V. (1982). Sequence and structural homologies among Type I and Type II interferons. *Nature* **300,** 379–81.

De Maeyer, E. and De Maeyer-Guignard, J. (1979). Considerations on mouse genes influencing interferon production and action. In, *Interferon 1979* (ed. I. Gresser) pp. 75–100. Academic Press, London.

Dijkema, R., Pouwels, P., de Reus, A., and Schellekens, H. (1984). Structure and expression in *E. coli* of a cloned rat interferon-α gene. *Nucleic Acids Res.* **12,** 1227–42.

Edge, M. D., Greene, A. R., Heathcliffe, G. R. Meacock, P. A., Schuch, W., Scanlon, D. B., Atkinson, T. C., Newton, C. R., and Markham, A. F. (1981). Total synthesis of a human leukocyte interferon gene. *Nature* **292,** 736–62.

Epstein, D. J. and Epstein, L. B. (1982). Genetic control of the response to interferon. *Texas Rep. Biol. Med.* **41,** 324–31.

Gillespie, D., Pequignot, E., and Carter, W. A. (1984). Evolution of interferon genes. In *Interferons and their applications.* (ed. P. E. Came and W. A. Carter) pp. 45–63, Berlin.

Goeddel, D. V., Leung, D. W., Dull, T. J., Gross, M., Lawn, R. M., McCandliss, R., Seeburg, P. H., Ullrich, A., Yelverton, E., and Gray, P. W. (1981). The structure of eight distinct cloned human leukocyte interferon cDNAs. *Nature* **290,** 20–6.

Gray, P. W., Leung, D.W., Pennica, D., Yelverton, E., Najarian, R., Simonson, C. C., Derynck, R., Sherwood, P. J., Wallace, D. M., Berger, S. L., Levinson, A. D., and Goeddel, D. V. (1982). Expression of human immune interferon cDNA in *E. coli* and monkey cells. *Nature* **295,** 503–7.

Gray, P. W. and Goeddel, D. V. (1983) Cloning and expression of murine immune interferon cDNA. *Proc. Nat. Acad. Sci. USA* **80,** 5842–6.

Hauser, H., Gross, G., Bruns, W., Hochkeppel, H. K., Mayr, H., and Collins, J. (1982). Inducibility of human-β IFN gene in mouse L-cell clones. *Nature* **297,** 650–4.

Higashi, Y., Sokawa, Y., Watanabe, Y., Kawada, Y., Ohno, S., Takoaka, C., and Taniguchi, T. (1983). Structure and expression of a cloned cDNA for mouse interferon-β. *J. Biol. Chem.* **258,** 9522–9.

Horisberger, M. A., Staeheli, P., and Haller, O. (1983). Interferon induces a unique protein in mouse cells bearing a gene for resistance to influenza virus. *Proc. Nat. Acad. Sci. USA* **80,** 1901–14.

Kingsman, S. M. and Kingsman, A. J. (1983). The production of interferon in bacteria and yeast. In, *Interferons, from molecular biology to clinical application. SGM Symposium 35,* (ed. D. C. Burke and A. G. Morris) pp. 211–54. Cambridge University Press, Cambridge.

Nagata, S., Mantei, N., and Weissmann, C. (1980) The structure of one of the eight or more distinct chromosomal genes for human interferon-α. *Nature* **287,** 401–8.

—— Brack, C., Henco, K., Schambock, A., and Weissmann, C. (1981). Partial mapping of ten genes of the human interferon-α family. *Interferon Res.* **1,** 333–6.

Pestka, S., Kelder, B., Familletti, P. C., Moschera, J. A., Crowl, R., and Kempner, E. S. (1983). Molecular weight of the functional unit of human leukocyte, fibroblast and immune interferons. *J. Biol. Chem.* **258,** 9706–9.

Raj, N. B. K. and Pitha, P. M. (1980). Synthesis of new proteins associated with the induction of interferon in human fibroblast cells. *Proc. Nat. Acad. Sci. USA* **77,** 4918–22.

Rubinstein, M., Rubinstein, S., Fomillette, P. L., Miller, R. S., Waldman, A. A., and Pestka, S. (1979). Human leukocyte interferon: production, purification to homogeneity and initial characterisation. *Proc. Nat. Acad. Sci. USA* **76,** 640–4.

Samantha, H., Yoshie, O., Schmidt, H., St-Laurant, G., Floyd-Smith, G., Jayaram, B. L., and Lengyel, P. (1983). Interferons as inducers of mRNA and protein synthesis. In *The biology of the interferon system 1983* (ed. E. De Maeyer and H. Schellekens) pp. 239–43. Elsevier, Amsterdam.

Sehgal, P. B. and Sagar, A. D. (1980). Heterogeneity of poly(I)-poly(C)-induced human fibroblast interferon mRNA species. *Nature* **288,** 95–7.

Shaw, G. B., Boll, W., Taira, H., Mantei, N., Lengyel, P., and Weissman, C. (1983). Structure and expression of cloned murine IFN-α genes. *Nucleic Acids Res.* **II,** 555–73.

Shire, S. J. (1983). pH-dependent polymerization of a human leukocyte interferon produced by recombinant deoxyribonucleic acid technology. *Biochemistry* **22,** 2664–71.

Streuli, M., Hall, W., Stewart, W. E. II, Nagata, S., and Weissmann, C. (1981). Target cell specificity of two species of human interferon-α produced in *E. coli* and of hybrid molecules derived from them. *Proc. Nat. Acad. Sci. USA* **78,** 2848–52.

Taniguchi, T., Mantei, N., Schwarzstein, M., Nagata, S., Muramatsu, M., and Weissmann, C. (1980). Human leukocyte and fibroblast interferons are structurally related. *Nature* **285,** 547–9.

Taniguchi, T., Sakai, M., Fujii-Kuriyama, Y., Muramatsu, M., Kobayashi, S., and Sudo, T. (1979). Construction and identification of a bacterial plasmid containing the human fibroblast interferon gene sequence. *Proc. Jap. Acad. Sci.* **855,** 464–9.

Van Damme, J., Billiau, A., De Ley, M., and De Somer, P. (1983). An interferon-like or interferon-inducing protein released by mitogen stimulated human leucocytes. *J. Gen. Virol.* **64,** 1819–22.

Weil, J., Epstein, C. S., Epstein, L. B., and Grossberg, S. E. (1983). A unique set of polypeptides is induced by γ interferon in addition to those induced in common with α and β interferons. *Nature* **301,** 437–9.

Weissenbach, J., Chernajovsky, Y., Zeevi, M., Schulman, L., Soreg, H., Nir, V., Wallach, D., Perricaudet, M., Tiollais, P., and Revel, M. (1980). Two interferon mRNAs in human fibroblast: *in vitro* translation and *E. coli* cloning studies. *Proc. Nat. Acad. Sci. USA* **77,** 7152–6.

Weissmann, C. (1981). The cloning of interferon and other mistakes. In, *Interferon 3*

(ed. I. Gresser) pp. 101–34. Academic Press, New York.
—— Nagata, S., Boll, W., Fountoulakis, M., Fujisawa, A., Fujisawa, J. I., Haynes, J., Henco, K., Mantei, N., Ragg, H., Schein, C., Schmid, J., Shaw, G., Streuli, M., Taira, H., Todokoro, K., and Weidle, U. (1982). Structure and expression of human alpha-interferon genes. ICN–UCLA Symposium (March 1982) Chemistry and biology of interferons: relationship to therapeutics (ed. T. C. Merigan and R. M. Friedman) Vol. 35, pp. 295–326.

Wetzel, R., Levine, H., Estell, D. A., Shire, S., Finer-Moore, J., Stroud, R. M., and Bewley, T. A. (1982). Structure–function studies on human alpha interferon. *Interferons* pp. 365–76. Academic Press, New York.

Wilson, V., Jeffreys, A. J., Barrie, P. A., Boseley, P. G., Slocombe, P. M., Easton, A., and Burke, D. C. (1983). A comparison of vertebrate interferon gene families detected by hybridization with human interferon DNA. *J. Mol. Biol.* **166,** 457–75.

2 Interferons and gene expression

Michael L. Riordan and Paula M. Pitha-Rowe

2.1. INTRODUCTION

Although much is known about the control of gene expression in prokaryotes, we are only just beginning to acquire an understanding of the mechanisms of gene regulation in eukaryotes. The extremely complex interplay of events at the genomic, nuclear, and cytoplasmic levels in eukaryotes has made the task of studying single genes or families of genes very difficult. Thus it is that many studies of eukaryotic gene regulation have focused on genes that can be induced with exogenous agents, allowing the experimenter to regulate expression and more precisely examine events before, during, and after transcription. With the advent of molecular cloning techniques and procedures for inducing genes of interest into heterologous cells, a new horizon has opened up for the study of eukaryotic gene regulation.

Of the relatively small but growing number of inducible genes to be investigated thus far (for a review, see Kessel and Khoury 1983), the interferon genes have perhaps been studied most intensively. Many features of the interferon system make it attractive as a means of studying regulation of gene expression. Biological assays permit the detection of vanishingly small quantities of interferon protein, and these assays can differentiate interferons on the basis of species specificity. Interferons are expressed by many different cell types within an organism, and are found in many different eukaryotic species. They constitute a family of related yet different genes which, depending on the inducer, cell type, or interferon subtype, may exhibit coordinate expression or appear to be regulated by different mechanisms. Because of the extensive homology within this family, it may be possible to attribute differences in the inducibility or expression of individual genes to specific nucleotide sequences or structural properties of the DNA within, or flanking, those genes. A final, very important feature is that interferon genes of all three types (α, β, and γ) have been cloned and their chromosomal genes have been isolated from human DNA libraries; thus the tools are in hand.

On the basis of antigenicity, the human and mouse interferons have been divided into three different groups, α, β, and γ (Stewart, Blalock, Burke, Chany, Dunnick, Falcoff, Friedman, Galasso, Joklik, Vilcek, Younger, and Zoon 1980). The physico-chemical properties of α and β-interferons, which are acid-stable proteins and can be induced by virus infection, differ from mitogen-induced γ-interferon, which is acid-labile. Sequence analysis of molecularly cloned human and mouse α-interferons has shown that these proteins are coded by a multiple gene family (Nagata, Mantei, and Weissmann 1980a; Goeddel, Leung, Dull, Gross, Lawn, McCandliss, Seeburg, Ullrich, Yelverton, and Gray 1981; Shaw, Boll, Taira, Montei, Lengyel, and Weissmann 1983; Kelley and Pitha, unpublished). The human α genes were found to comprise at least 13 non-allelic authentic genes, nine allelic genes and six pseudogenes. In contrast, there seems to be only one well characterized human (Taniguchi, Guarente, Roberts, Kimelman, Douhan, and Ptashner 1980) and mouse (Higashi, Sokawa, Watanabe, Kawade, Ohno, Takaoka, and Taniguchi 1983) β-interferon gene, and although the presence of other β-interferon genes has been postulated in both man and mouse, their existence needs further verification. (For reviews, see Fiers, Remaut, Devos, Cheroutre, Cantreras, Gheysen, Degrave, Stanssens, Tavernier, Taya, and Content 1982; Weissmann, Nagata, Boll, Fountoulakis, Fujisawa, Fujisawa, Haynes, Henco, Mantei, Ragg, Schein, Schmid, Shaw, Streuli, Taira, Todokoro, and Weidle 1982; Stewart 1979.)

Both α and β-interferon genes are unspliced (i.e. do not contain introns) and are localized in humans on the short arm of chromosome 9 (Shows, Sakaguchi, Naylor, Goeddel, and Lawn 1982) and in mice on chromosome 4 (Kelley, Kozak, Dandoy, Sor, Skup, Windass, De Maeyer-Guignard, Pitha, and De Maeyer 1983), in close proximity to the H-15 minor histocompatability locus. Some of the α-interferon genes are closely linked; a cluster of three α-interferon genes, separated by 12.3 kilobase (kb) and 5 kb noncoding DNA sequences, was identified in human DNA, and in mice a cluster of five different α-interferon genes was identified recently by Kelley and Pitha (unpublished) in a 300kb fragment of mouse BALB/c DNA.

The first 200 nucleotides of the 5′ flanking region of both human and mouse α-interferon genes show a striking similarity (up to 75 per cent homology; Goeddel *et al.*, 1981; Kelley and Pitha, unpublished). Segments of extensive homology have been identified in the 5′ (upstream) region of all human and mouse α and β-interferons so far examined, and it is attractive to speculate that these contain consensus sequences regulating the expression and inducibility of these genes. The sequences corresponding to the Hogness-Goldberg box, which is believed to play an important role in the initiation of transcription, was found 31 nucleotides before the transcription initiation (cap) site of both α and β mRNAs. In the majority of the human α-interferon genes, the sequence found is TATTTAA, while the sequence found in front of the β-interferon gene is TATAAATA. Alteration of TATA sequences in other genes has been shown to alter their specificity and level of transcription, but whether the alteration of TATA box effects the efficiency of the transcription of α-interferon genes is not yet known with certainty.

The sequence of the 3′ noncoding region of α-interferon genes shows a larger degree of variability than the sequences of the coding region or the 5′ nontranslated region. The 3′ flanking region of the β-interferon gene shows two interesting features. The first is the restriction site homologies that were found to be preserved 2–3kb downstream of the β-interferon gene in genomes of a variety of primates (Wilson, Jeffreys, Barrie, Boseley, Slocombe, Easton, and Burke 1983). The conservation of these regions through evolution may suggest a functional role for these sequences. This suggestion is further supported by the observation that in human DNA another gene inducible by poly rI.rC together with β-interferon was found to be localized in this region (Gross, Mayr, and Collins 1981). This gene, which codes for 9S RNA in induced human fibroblasts, does not seem to be expressed in Namalwa cells induced with Sendai virus to synthesize interferon. It is not yet clear what the relation is of this gene to the expression of β-interferon genes. However, these findings suggest that both the 5′ and 3′ flanking sequences of the interferon genes may have some functional role.

Several experimental techniques have been fundamental to studies of the molecular genetics of the interferon system, and some of them will be described before proceeding to discussions of the characteristics of interferon expression and its control; these discussions will focus on α and β-interferon rather than γ-interferon, as the latter is a relative newcomer and not nearly as well studied.

2.2. EXPERIMENTAL METHODS

There are two specific features of interferon polypeptides that make their detection and assay unique. One is their high biological activity, which made it possible to detect picogram quantities of interferons long before radioimmunoassays for interferons were available. This sensitive biological assay, based on the inhibition of virus replication in interferon-treated cells, is ten- to a hundred-fold more sensitive than radioimmunoassay using either monoclonal or polyclonal antibodies. The second feature is the species specificity of the majority of interferons; that is, they are much more active on homologous than heterologous cells. The first characterization and isolation of interferon mRNA was based on these two features.

2.2.1. Isolation of interferon mRNA

DeMaeyer-Guignard, De Maeyer, and Montagnier (1972) showed the existence of mouse interferon mRNA by translating the total mRNA population isolated from L cells induced by Newcastle Disease virus (NDV) to synthesize interferon. This was accomplished by introducing the mRNA into chick cells. The medium from the chick cells that had taken up this RNA in the presence of DEAE dextran, contained antiviral activity that was specific for mouse cells, indicating that this activity was a translation product of the mouse interferon mRNA. Interferon synthesized by chick cells alone would have had antiviral activity on chick, but not on mouse cells. Similarly, Reynolds and Pitha (1974) were able to translate, in chick cells, human interferon mRNA from human fibroblasts induced by double-stranded RNA (dsRNA), and by using this assay they were able to compare the kinetics of the synthesis of interferon and its mRNA. The main disadvantages of this assay, although used successfully by several other laboratories, were a large quantitative variation from experiment to experiment and an inability to sufficiently radiolabel the synthesized interferons because of the large internal pool of amino acids in eukaryotic cells.

In order to label the newly synthesized peptides coded by the injected mRNA sequences, Reynolds, Premukar, and Pitha (1975), microinjected total mRNA from induced fibroblasts into the cytoplasm of *Xenopus* oocytes. This system, introduced by Gurdon, Lane, Woodland, and Martaix (1971), had been shown previously to be very useful for the synthesis of labelled proteins coded by microinjected mRNA. What Reynolds and colleagues showed is that not only did the labelled interferon synthesized in oocytes have the same size and physicochemical properties as mature interferon synthesized in human fibroblasts, but it was also biologically active. The amounts of biologically active interferons synthesized in *Xenopus* oocytes were much higher than in any *in vitro* systems used, and the assay was very reproducible. This oocyte assay was used for the characterization and purification of interferon mRNAs, for studies of regulation of interferon gene expression in induced cells, and for the detection of cloned interferon cDNAs.

2.2.2. Isolation of interferon cDNA clones

Clones carrying interferon cDNA inserts were identified and isolated on the principle that only DNA from a clone carrying the interferon sequence would hybridize to interferon mRNA, which after elution could be identified by translation in *Xenopus* oocytes or in an *in vitro* system. Thus DNA was isolated from pooled clones and hybridized to mRNA from induced cells containing interferon mRNA. Pools of clones containing interferon cDNA sequences were further divided and individual clones containing interferon cDNA identified (Derynck, Content, De Clercq, Volckaert, Tavernier, Devos, and Fiers 1980; Nagata, Taira, Hall, Johnsrud, Streuli, Escodi, Boll, Cantell, and Weissmann 1980b). This approach, although laborious, was fruitful in several laboratories and opened the way for cloning of any protein for which a sensitive biological assay is available (e.g. α, β, γ-interferons, interleukin-2, interleukin-3).

2.3. REGULATION OF INTERFERON SYNTHESIS IN HOMOLOGOUS CELLS

The synthesis of type 1 interferons (α and β) includes a number of limiting steps. Thus the levels of interferon synthesized are determined by the inducer, efficiency of transcription of interferon genes after induction, and by the stability and translational efficiency of the interferon mRNA. (For a review, see Burke 1982). Since interferon is an extracellular protein, additional regulations may be connected with processing of a pro-interferon peptide into interferon and its release from the cell. In the following section, we shall summarize our knowledge about these processes.

2.3.1. Inducers

Both viruses and dsRNA can efficiently induce type 1 interferons *in vivo* and *in vitro*. The level of induction depends on the inducer and shows cell type specificity. Thus the α-interferon genes seems to be most efficiently induced in B and null lymphocytes and macrophages whereas the β-interferon gene is induced in fibroblast and epithelial cells. (Baron, Dianzani, and Stanton 1982). Not all viruses induce interferon with the same efficiency. The most effective interferon inducers are the RNA viruses of the myxovirus and paramyxovirus groups. All experimental evidence indicates that the inducing entity is ds viral RNA which is formed (even transiently) during the viral replicative cycle of both RNA and DNA viruses (Marcus and Sekellick 1980). Thus, for example, U.V.-inactivated NDV is able to induce interferon as long as inactivation does not abolish completely the function of RNA replicase and at least part of the viral genome is transcribed. The mutant strains of inducing viruses which show restriction in cell entry and consequently, therefore, in replication are generally much poorer interferon inducers (Kelley and Pitha, unpublished). The most direct evidence for the role of viral dsRNA in viral induction comes from the work of Marcus, who was able to demonstrate that a single particle of a vesicular stomatitis virus (VSV) mutant consisting of dsRNA of both VSV complementary RNA strands can induce interferon production in a cell (Marcus and Sekellick 1977). Whether each virus induces a similar set of interferon genes and the diversity in expression is given by the cell type, or whether there is a virus-associated specificity in the induction of type 1 genes remains to be determined. Raj and Pitha (1983), however, have shown that at least in primary human fibroblasts, both virus (NDV) and polyriboinosinic acid.polyribocytidilic acid (poly rI.rC) induce only one type of interferon gene, indicating that the cell type, rather than the virus, may play the greater role in determining which type of interferon gene is induced. The question of whether all induced interferon genes are coordinately expressed after induction is also of interest, and recent work of Shuttleworth, Morser, and Burke (1983), Raj, Kellum, Kelley, and Pitha (1983), and Kelley and Pitha (unpublished results), indicate that this may be the case. In Namalwa cells (a human lymphoblastoid line) induced with Sendai virus, both α and β-interferon mRNAs were shown to be synthesized within the same time interval after induction. In mouse L cells induced with NDV, several α-interferon genes are induced with the same kinetics.

It should be mentioned at this point that current views of the mechanism of interferon induction by poly rI.rC involve some controversial points, mainly centring on the question of whether the cellular uptake of poly rI.rC is required for induction, or whether poly rI.rC, like interferon, exerts its effect through interaction with the plasma membrane (Pitha 1981). Although none of the experimental approaches so far attempted has definitively solved this question, it is our opinion that the present evidence suggests that poly rI.rC has to be taken up by the cell in order to induce interferon synthesis. Treatments that enhance the uptake of nucleic acid into cells (DEAE dextran, polyamine, $CaCl_2$ precipitation) increase the inducing potential of poly rI.rC. The inducibility is also dependent on the molecular weight of poly rI.rC, another factor that can affect the uptake of this molecule. Microinjection of poly rI.rC into inducible cells followed by assay for interferon mRNA at the single cell level would help answer this question.

Thus the first limiting step in interferon induction is the production of dsRNA during the viral replication cycle or uptake of dsRNA into the cells. Nothing is known about the quantitative aspect of this process, but all the experimental data using non-replicative viruses indicate that one or only a few virus particles per cell are sufficient for interferon induction. It should also be mentioned at this point that although all cells are capable of producing interferon, the induction process is somehow related to the physiological state of the cell or to the cell cycle. Thus even with a clonal population of cells subjected *in vitro* to viral infection of high multiplicity (many more than one virus particle per cell, resulting in virtually every cell being infected), interferon is not synthesized by 100 per cent of the treated cells. The levels of interferon synthesis are higher in nondividing or quiescent cells than in growing or rapidly dividing cells; whether this phenomenon reflects dependence of interferon gene transcription on the cell cycle or post-transcriptional factors, such as the availability of ribosomes for the translation of interferon mRNA in nondividing cells or a higher stability of interferon mRNA in the absence of cellular protein synthesis (see discussion later), remains to be seen.

2.3.2. Transcription

Although we have quite good evidence that dsRNA is the entity that induces or activates the transcription of Type 1 interferon genes, the molecular mechanisms by which these genes are activated are not clear.

With only a few exceptions, such as cancer cells or lymphoblastoid cell lines, interferon synthesis does not occur spontaneously in uninduced cells; no transcripts of interferon genes are detected in isolated nuclei and no interferon mRNA is found in uninduced cells (Raj *et al.* 1983). In human fibroblasts induced with poly rI.rC, transcription of the β-interferon gene does not require new protein synthesis, as it is transcribed even when cellular protein synthesis is inhibited by cycloheximide treatment (Raj and Pitha 1981). Thus in this system, the induction by dsRNA does not proceed through the induction of a *de novo* synthesized regulatory protein which would activate or derepress transcription of β-interferon gene. Since under certain conditions both α and β-interferon genes seem to be activated coordinately, we can assume that the same may be true for induction of the α-interferon genes.

As with other eukaryotic genes, the transcription of interferon genes in homologous systems (e.g. human genes in human cells) starts at a specific cap site; no transcript initiated in the upstream region of α or β genes has been detected so far in human cells. An interesting difference between the relative rates of transcription of α and β-interferon genes was observed by using nuclei isolated from Namalwa cells induced with Sendai virus to produce interferon (Raj *et al.* 1983). It has been shown that in Namalwa cells, several α-interferons and one β-interferon are synthesized. If all these genes are transcribed at comparable rates, one would expect to find the relative levels of α transcripts to be several times higher than the level of the β transcripts. The results of Raj *et al.* (1983) and Shuttleworth *et al.* (1983), however, indicate that the levels of α and β transcripts are comparable in spite of the fact that the α transcripts represent transcription of several α-interferon genes (as mentioned before, α genes have very similar DNA sequences and, therefore, one α probe detects a family of all α genes). This indicates that the α and β-interferon genes are transcribed with different rates and efficiencies.

The termination of the transcription of α and β-interferon genes may also show an interesting difference. It was shown by Content, De Wit, Tavernier, and Fiers (1983) and confirmed in other laboratories (Seghal, May, La Forge, and Inouye 1982; Raj and Pitha, unpublished); that in fibroblasts induced with NDV or with poly rI.rC in the presence of cycloheximide, in addition to β mRNA (11S), low levels of long transcripts (24S) (10–100 fold less than β mRNA) can be detected by hybridization with a β cDNA probe. These 24S transcripts were shown to be initiated in a similar position to β mRNA, but were not terminated at the poly(A) site; they represented a read-through transcription into the 3′ flanking region. It is not yet clear whether the 24S RNAs represent primary transcripts of the β-interferon gene which are then processed into 11S β mRNA or whether only a small percentage of β-interferon gene transcript is not properly terminated and instead transcribed beyond the poly(A) site. The 24S transcript cannot be detected in fibroblast cells induced with poly rI.rC in the absence of cycloheximide, and thus it can not be excluded that they are formed only under conditions when cellular protein synthesis is inhibited (cellular protein synthesis is also inhibited by infection with NDV).

Long transcripts hybridizing with an α cDNA probe have been detected only in one laboratory and have not yet been confirmed by others. Using such a probe, Seghal *et al.* (1982) and May, Seghal, La Forge, and Inouye

(1983) detected, in 5′-bromodeoxyuridine (BrdU)-treated Namalwa cells induced with Sendai virus, several long mRNAs in addition to the 12S α mRNA. The sizes and relative levels of these long transcripts seem, however, to differ among different reports. These observations, although very interesting, have to be considered only tentative as yet. The hybridization conditions under which these transcripts were detected included a cDNA probe containing the 3′ untranslated region and nonstringent washing; misleading information in other systems has been recently obtained using similar approaches (Patient 1984). Using the same cells and methods of induction, other laboratories do not seem to be able to detect the long α mRNA transcripts. Shuttleworth *et al.* (1983) have shown that BrdU treatment of Namalwa cells before induction increases the steady-state levels of α mRNA (12S) in these cells without altering the nature of the transcripts. Raj *et al.* (1983) examined the transcription of the A(α_2) interferon gene in Namalwa cells using isolated nuclei and found that the transcription of this gene is terminated in the vicinity of the poly(A) site and does not proceed further into the 3′ flanking region. Accordingly, no transcripts larger than 12S were detected in the induced cells either by Northern hybridization or S1 mapping.

The transcription of interferon genes shows cell type specificity. In primary human fibroblast cells, poly rI.rC induces the transcription of the β-interferon gene exclusively (Raj and Pitha 1983), whereas in other fibroblasts, trisomic for chromosome 21, the transcription of α-interferon genes was detected; however the levels of α mRNA were at least 100-fold lower than those of β mRNA (Raj and Pitha 1983; Hayes, Yip, and Vilcek 1979). In lymphocytes, lymphobastoid cell lines and promyelocyte cell lines both α and β-interferon genes can be induced (Havell, Yip, and Vilcek 1977). This may indicate the requirement of trans (cellular) factors for the induction of α-interferon genes that are present in the cells of lymphoid origin.

It has been shown by a number of laboratories that treatment of the cells with BrdU or sodium butyrate before the induction increases the amount of interferon synthesized in these cells (e.g. Adolf and Swetly 1981; Shuttleworth, Morser, and Burke 1982). The results of Adolf and Swetly (1982) indicated that in different lymphoblastoid lines, butyrate treatment preferentially increased synthesis of α-interferons while the levels of β-interferons were not greatly altered. Shuttleworth *et al.* (1983), however, found when they assayed hybridizable α mRNA that butyrate and BrdU increased the relative levels of both α and β mRNA present in the cells. Raj and Pitha (unpublished results) labelled with S^{35}-methionine, interferon synthesized in induced Namalwa cells with and without prior butyrate treatment, and assayed it by SDS gel electrophoresis after immunoprecipitation with the appropriate antiserum. They found that the synthesis of both α and β-interferons was enhanced in butyrate-treated cells by approximately 10-fold when compared to the untreated cells. All of these experiments provide strong evidence that enhancement of interferon synthesis by butyrate and

BrdU may be primarily due to the enhancement of transcription of the interferon genes.

2.3.3. Post-transcriptional regulation

The synthesis of β-interferon in human fibroblasts induced with poly rI.rC is a transient phenomenon; interferon is synthesized within a couple of hours of induction, but in a few hours its synthesis is rapidly turned off. If the interferon induction is carried out in the absence of cellular protein synthesis and RNA synthesis is stopped a few hours after induction by actinomycin D treatment, then interferon synthesis is greatly enhanced and continues for many hours. This effect was originally described by Tan, Armstrong, Ke, and Ho (1970) and later worked out in detail by Vilcek's group (Vilcek, Havell, and Kohase 1976). It was shown by several groups (e.g. Cavalieri, Havell, Vilcek, and Pestka 1977; Sehgal, Leyles, and Tamm 1978; Raj and Pitha 1977) that the switch-off in interferon synthesis is accompanied by a decrease in the levels of translatable interferon mRNA, and Raj and Pitha (1981) have shown that during the shut-off period the β mRNA sequences are rapidly degraded with a half-life of approximately 30 min. The breakdown in interferon mRNA was shown to be coupled to cellular protein synthesis, and thus the factor which regulates the β-interferon synthesis is a protein. To determine whether the short appearance of interferon β mRNA in the cells is due to the transient activation or derepression of the β-interferon gene, the relative rate of transcription was measured in nuclei isolated from the cells at different times post-induction (Raj and Pitha 1983). The results show that the β-interferon gene was transcribed during the shut-off period of β-interferon synthesis when no interferon mRNA could be detected in the cells. The fact that no β mRNA sequences were detected in the cells at times when the β-interferon gene was actively transcribed indicates that the protein regulating β mRNA levels acts at the post-transcriptional level; it is obvious, however, that the degradation is specific for β mRNA since the half-life of the total poly(A) mRNA in these cells is not greatly affected. These results indicate that the synthesis of β-interferon in poly rI.rC induced human fibroblasts is controlled both by the activation of interferon gene transcription and alteration of the β-interferon mRNA stability.

2.4. EXPRESSION IN HETEROLOGOUS CELLS

Several recently acquired methods for introducing human genes into cultured mammalian cells have been used to study the regulation of expression of interferon genes. These methods differ in the types of DNA vectors used and consequently in the state in which the exogenous DNA is retained within the cell, i.e., whether it is integrated into the cellular genome or maintained episomally, and whether it is stably or only transiently maintained within the recipient cell line. Thus interferon genes and other sequences of interest can be stably integrated into host chromosomal DNA by cotransformation with biochemically selectable markers, such as the herpes simplex virus thymidine kinase gene (HSV-TK) or alternatively, genes can be linked to SV40-derived vectors that are retained extrachromosomally, and only expressed transiently, as their residence in the cell is limited by the vector's lytic cycle. An additional, very promising vector is a derivative of the bovine papilloma virus (BPV) that can morphologically transform mouse cells and is maintained as a stable, multicopy episome.

Several groups have used these methods to try to determine what portions of the human α and β-interferon genes, including their flanking sequences, are required for induction, and to shed light on the question of whether the initiation of transcription or a post-transcriptional factor such as stabilization of mRNA plays the major role in induction. Such studies are facilitated by a number of fortuitous features of the interferon system, such as the presence of interferon genes in the commonly used host cells (usually mouse), which are induced by the same agents (NDV or poly rI.rC) that are used to induce the exogenous human genes, thus permitting an internal control for the efficacy of inducer treatment. The ability to distinguish between host interferon and interferon encoded by the exogenous genes is of course critical in this respect. Likewise, it is fortuitous that the homology of host and exogenous interferon genes (e.g., mouse and human) is low enough that hybridization with appropriate probes, using proper wash conditions, can discriminate between the two sets of genes and their transcripts.

The findings of a number of groups investigating interferon expression in heterologous cells are described below, and they are summarized in a simple format in Table 2.1.

2.4.1. Studies of human *β*-interferon (HuIFN-*β*)

Hauser, Gross, Bruns, Hochkeppel, Mayr, and Collins (1982), using poly rI.rC as an inducer, demonstrated the inducibility of a 36kb segment of human DNA containing the IFN gene (the translated sequence of which is 561 nucleotides long) after it had been cotransferred with the HSV-TK gene into mouse LTK-cells. This segment contained not only the HuIFN-β gene, but also several other genes which could be induced together with the β-interferon gene. A smaller, 1.9kb fragment containing only the HuIFN-β gene and flanking sequences, when transfected into the mouse cells, was inducible as well, indicating that the adjacent co-induced genes were not essential for HuIFN-β induction.

Pitha, Cuifo, Kellum, Raj, Reyes, and Hayward (1982) also used the HSV-TK gene to cotransfer HuIFN gene sequences into mouse LTK-cells. One of the sequences contained an 840 base-pair (bp) segment of HuIFN-β cDNA spanning the entire coding region but lacking the 5′ flanking region and 15 nucleotides of 5′ untranslated transcript sequence. A second construct used for transfection contained only 560bp from the HuIFN-β coding region,

Table 2.1. *Induced expression of cloned human interferon genes introduced into mammalian cells*[a]

IFN gene	bp of IFN gene sequence 5′ of cap site	Coding sequence	Means of transfection	Integrated/ nonintegrated	Host cell type	Inducer	Reference
β	283	HuIFN-β	cotransfer with HSV-TK	integrated	mouse LTK^-	poly rI.rC, NDV	Hauser *et al.* (1982)
β	0	HuIFN-β	cotransfer with HSV-TK	integrated	mouse LTK^-	poly rI.rC	Pitha *et al.* (1982)
β	283	HuIFN-β	cotransfer with neo	integrated	mouse L, rabbit RK13	poly rI.rC	Canaani and Berg (1982)
β	283	HuIFN-β	cotransfer with gpt	integrated	mouse FM3	poly rI.rC, NDV	Ohno and Taniguchi (1982)
β	283 (−283 to +20)	HSV-TK	contransfer with gpt	integrated	mouse FM3	NDV	Ohno and Taniguchi (1983)
β	243	HuIFN-β	cotransfer with DHFR	integrated, amplified	Ch. hamster ovary	poly rI.rC, NDV	McCormick *et al.* (1984)
β	286 [−186 to −142][b]	HuIFN-β	SV40 vector	nonintegrated	monkey AP-8	poly rI.rC	Tavernier *et al.* (1983)
β	285	HuIFN-β	SV40 vector	nonintegrated	monkey CV-1	poly rI.rC	Maroteaux *et al.* (1983)
β	40	HuIFN-β	SV40 vector	nonintegrated	monkey CV-1	poly rI.rC	Maroteaux *et al.* (1983)
β	350	HuIFN-β	BPV vector	nonintegrated	mouse C127	poly rI.rC	Zinn *et al.* (1982)
β	210 [−77 to −19][b]	HuIFN-β	BPV vector	nonintegrated	mouse C127	poly rI.rC	Zinn *et al.* (1982)
β	350	HuIFN-β	BPV vector	nonintegrated	mouse C127	poly rI.rC, NDV	Mitrani-Rosenbaum *et al.* (1983)
α_1	5400	HuIFN-α_1	cotransfer with HSV-TK	integrated	mouse LTK^-	NDV	Mantei and Weissman (1982)
α_1	670 (∼−670 to −6)	rabbit β-globin	cotransfer with HSV-TK	integrated	mouse LTK^-	NDV	Weidle and Weissmann (1983)
α_1	117	HuIFN-α_1	cotransfer with HSV-TK	integrated	mouse LTK^-	NDV	Ragg and Weissmann (1983)

[a]See text for explanations.
[b]Sequences found by deletion studies to be required for induction.

inserted under the transcriptional control of the TK promoter. Several of these transformants were found to produce HuIFN-β (HuIFN-β mRNA was not measured) after induction with poly rI.rC and cycloheximide, findings that contrast with several other studies indicating that 5′ flanking sequences are required for induction.

Canaani and Berg (1982) introduced into mouse L and rabbit kidney cells either of two different SV40 hybrid plasmid vectors containing the HuIFN-β gene and, as a selectable marker, the bacterial phosphotransferase gene (neo). One plasmid contained the HuIFN-β transcribed region with both 5′ and 3′ flanking sequences, of 283 and 784 base pairs, respectively. In the other plasmid, the 3′ flanking region was replaced with an SV40 DNA segment containing the small tumour antigen intervening sequence and early region polyadenylation signal. Induction of HuIFN-β and its mRNA by poly rI.rC was demonstrated in transformants of either plasmid, indicating that the 3′ flanking sequence is not essential for induction and that the transcribed region and its 283bp 5′ flanking sequence contain sufficient information for induction. The authors noted that even in transformants found to have the HuIFN-β and neo genes arranged in tandem, as in the original transducing plasmid, expression of the neo gene was not increased by poly rI.rC induction, suggesting that induction is localized to the IFN gene and does not result in activation of an extended region of DNA. This, in turn, implies that the genes found to be concomitantly induced with interferon [noted above in the transfections performed by Hauser *et al.* 1982 and found by others in human fibroblasts (Raj and Pitha 1980; Gross *et al.* 1981)] may have their own induction-responsive sequences.

Ohno and Taniguchi (1982), who previously had also demonstrated inducibility of a 1.8kb chromosomal DNA segment containing the HuIFN-β gene that had been introduced into mouse FM3A cells, subsequently studied the inducibility of the isolated HuIFN-β 5′ flanking sequences (Ohno and Taniguchi 1983). They inserted this sequence (−284 to +20, relative to the presumed transcription initiation site) in front of the HSV-TK structural gene in a plasmid vector (pSV2-ecogpt) containing the *E. coli* guanine phosphoribosyl transferase gene as a selectable marker. Mouse FM3A cells transformed by this plasmid produced TK-specific transcripts with a 5′ terminus corresponding to that of HuIFN-β when treated with NDV. Thus the 5′ flanking sequence alone was found to respond to the HuIFN-β induction signal.

Further evidence that the 5′ flanking region is required for HuIFN-β induction was found by McCormick, Trahey, Innis, Diekmann, and Ringold (1984), who transfected dihydrofolate reductase (DHFR)-negative Chinese hamster ovary cells with plasmid DNA containing the HuIFN-β gene and mouse DHFR cDNA. They were able to obtain transformants containing amplified copies of both the DHFR and HuIFN genes by exposing cells to increasing methotrexate concentrations. In those cells transfected with plasmid containing a 1.8kb fragment of human DNA containing interferon-β coding and flanking sequences (243bp of 5′ sequences and 714bp of 3′ sequences), treatment with poly rI.rC in the presence of cycloheximide (superinduction) or with NDV, resulted in expression of high levels of HuIFN-β (up to 10^{10} i.u. l^{-1} medium). When the 5′ flanking sequences in the transfecting plasmid were replaced with the SV40 early promoter, HuIFN-β was expressed constitutively at low levels but was not inducible with poly rI.rC or NDV.

All of the examples of heterologous HuIFN-β expression described so far involved exogenous DNA that was integrated into chromosomal DNA. An alternative approach, to introduce heterologous IFN genes that are maintained extrachromosomally and are thus not influenced by surrounding host DNA sequences and chromatin, has been used by Tavernier, Gheysen, Duerinck, and Van der Heyden (1983) who transfected monkey AP-8 cells with transiently maintained SV40 vectors into which the HuIFN-β gene, with or without its 5′ flanking sequence, had been inserted. On treatment with poly rI.rC, no induction was observed in cells transfected with the HuIFN-β gene lacking the 5′ flanking region, whereas cells transfected with the HuIFN-β gene including its 286bp 5′ flanking sequence were inducible. (Interferon, rather than HuIFN-β transcripts, was measured in these studies.) Furthermore, by deleting progressively longer DNA segments starting at the 5′ terminus of the 5′ flanking region, they found a drop in inducibility when sequences were deleted between nucleotides −186 and −142 (numbered relative to the putative cap site). The authors noted that this region contains a 19bp sequence that has an intriguing homology with a consensus sequence found in the 5′ flanking region of steroid hormone responsive genes, to which the progesterone–receptor complex may bind (Mulvihill, Le Pennec, and Chambon 1982).

In a study using a similar short-term, episomal SV40 transfection vector in monkey kidney CV-1 cells, Maroteaux, Kahana, Mory, Groner, and Revel (1983) also introduced the HuIFN-β gene with flanking sequences (~285bp 5′ to the presumed cap site and 700bp 3′ to the polyadenylation site), and found it to be inducible with poly rI.rC, as measured by IFN biological assay and transcript detection. In contrast to the findings of Tavernier *et al.* (1983), mentioned above, when these authors deleted 5′ flanking sequences up to 40bp before the presumed cap site of the HuIFN-β gene, they could still demonstrate inducible production of properly initiated HuIFN-β transcripts. However, the inducibility and level of expression of this 5′-deleted gene varied considerably depending on its location and orientation within the vector.

The third type of vector used in transfection experiments with the HuIFN-β gene was a plasmid containing a portion of the bovine papilloma virus (BPV). Zinn, Mellon, Ptashne, and Maniatis (1982) found that in C127 mouse fibroblasts transformed with a BPV-HuIFN-β recombinant containing HuIFN-β flanking sequences (350 and 660bp at the 5′ and 3′ termini, respectively), HuIFN-β mRNA and protein could be induced by poly rI.rC. In a later study, Zinn, Di Maio, and Maniatis (1983) analyzed the expression of deletion mutants of the HuIFN-β gene, again introduced into mouse fibroblasts on a BPV vector. In cell lines transformed by a BPV-IFN plasmid containing the HuIFN-β coding sequences with 210bp 5′ to the cap site, HuIFN-β mRNA was induced ~400-fold by poly rI.rC. Deletion studies of the 5′ flanking sequence implicated two regions as being important in regulation. The −77 to −19 region (numbered relative to the cap site) was found to be required for constitutive and induced IFN gene expression, both of which were greatly reduced by deletion to −73. When sequences between −210 and −107 were deleted, constitutive expression increased 5- to 10-fold and the induced expression was essentially unchanged. These data suggested that the region around −77 demarcates the 5′ border of an inducible promoter, and that the −210 to −107 and −77 to −19 regions are distinct regulatory sequences.

Although deletion of sequences between −210 and −107 did not significantly alter the level of induced expression, it did alter the kinetics of mRNA and IFN production. Deletions down to −95 or −77 caused maximal levels of HuIFN activity and mRNA to be reached 2–4 hr sooner after poly rI.rC induction than in cells with the full −210 flanking sequence. The authors, considering this shift in kinetics to be similar to that caused by 'priming' (treating certain cell types with homologous IFN before induction), speculated that priming with IFN derepresses a control region between −210 and −107.

The utility of BPV-derived vectors in heterologous expression systems was also demonstrated by Mitrani-Rosenbaum, Maroteaux, Mory, Revel, and Howley (1983), who transformed mouse C127 cells with a BPV-derived plasmid containing a 1.6kb segment of the genomic HuIFN-β gene. Transformed cells constitutively produced low levels of HuIFN-β and responded to induction with either poly rI.rC or inactivated NDV. The authors also verified that the heterologous genes were maintained episomally, rather than integrated into the cellular genome.

2.4.2. Studies of human α-interferon

Considerably fewer studies have been made of heterologously expressed human alpha interferon (HuIFN-α) genes. Mantei and Weissmann (1982) transfected mouse LTK-cells with a cloned HuIFN-α_1 gene (that had 5.4kb of 5′ flanking and 1.2kb of 3′ flanking sequences) linked to the HSV-TK gene as a selectable marker. They obtained cell lines that produced correctly initiated HuIFN-α_1 mRNA after induction with NDV, but not in non-induced, normal growth conditions. The authors also observed that the time course of the rise and fall of HuIFN-α_1 mRNA in the transfected cells after induction was very similar to that of the mouse IFN mRNA in the same cells.

This suggested that not only is the HuIFN-α_1 gene capable of responding to the mouse cell's mediation of the induction signal, but that it is also subject to the mouse cell's interferon shut-off mechanism. Combining these data with the findings of Weidle and Weissmann (1983) described above, they concluded that the region from -117 to -6 contains the sequences necessary for HuIFN-α_1 induction.

These authors also observed that deletion of all but 117bp of the 5′ flanking region resulted in a low level of correctly initiated transcripts even in non-induced cells. Since such transcripts were not found in cells containing the full 5′ flanking region, it was suggested that the sequences normally present upstream of -117 may have a repressive effect on transcription – a proposition not dissimilar to that made by Zinn *et al.* (1983), who suggested that a region between -210 and -107 of the HuIFN-β gene may exert a repressive effect on transcription.

2.5. IMPLICATIONS OF THESE STUDIES

There are numerous discrepancies in these findings from studies of expression of heterologous interferon genes in mammalian cells, perhaps in part attributable to differences in experimental protocols and expression vectors used. For example, heterologous genes that are chromosomally integrated – as occurred in several of these transfection experiments – integrate at different locations within the genome and thus may be subject to different influences from surrounding sequences and chromatin; likewise, differences in the relative position and orientation of interferon genes and BPV or SV40 promoter regions within the episomal vectors may have contributed to differences in inducibility. Furthermore, some investigators measured interferon protein rather than interferon mRNA, and there were differences in induction protocols, some of which used, in addition to poly rI.rC, the protein synthesis inhibitor cycloheximide, which is itself known to induce interferon synthesis, but very probably not by the same mechanism as dsRNA.

On balance, the evidence from these heterologous expression experiments so far suggests that induction of both α and β species is due more to an activation of transcription than to an increased stabilization of a rapidly turned-over IFN RNA. Responsiveness of HuIFN-α and -β genes to induction is probably conferred by their 5′ flanking sequences, perhaps by a sequence of as few as 117bp (in the case of HuIFN-α_1) or 77bp (in the case of HuIFN-β) 5′ to the transcription initiation site. Although these sequences are, by themselves capable of responding to the normal induction signals for interferon, other sequences, both upstream (to at least -210) and downstream (5′ untranslated sequences, coding region, and 3′ flanking sequences) may also play a role in modulating the induction response.

In heterologous expression experiments analogous to those used for

HuIFN genes, several other inducible genes have been shown to have inducer-responsive 5′ flanking regions (for a review see Kessel and Khoury 1983). Among them are the murine metallothionein I gene, induced by cadmium; *Drosophila* heat shock (Hsp 70) gene, induced by heat; mouse mammary tumor virus, responsive to dexamethasone; human growth hormone gene, also induced by dexamethasone; and ovalbumin gene, responsive to progesterone. For the murine metallothionein I promoter (Brinster, Chen, Warren, Sarthy, and Palmiter 1982) and the *Drosophila* heat shock promoter (Pelham 1982), 90 and 66bp of 5′ flanking sequence, respectively, were found to be sufficient for induction. The responsiveness of the ovalbumin gene to progesterone induction appeared to be mediated by a region between −222 and −95. (Dean, Knoll, Riser, and O'Malley 1983). Also Karin, Haslinger, Hottgreve, Richards, Krauter, Westphal, and Beato (1984) report that the metal ion responsive element in the human metallothionein-II_A (hMT-II_A) gene is a highly conserved dodecameric sequence present twice in the 5′ flanking region of the gene, at −38 to −50 and −138 to −150; in addition, induction of the hMT-II_A gene by glucocorticoids was found to require a separate control element, between nucleotides −237 and −268.

As mentioned in the introduction, there are regions of marked homology between the HuIFN-α and -β genes in their 5′ flanking regions, and it is enticing to correlate conserved regions of homology with regions that have now been implicated as having regulatory functions. It has been pointed out, for example, that a purine-rich region between roughly −100 and −60, lying within the region found to be essential for induction of HuIFN-α_1, is highly conserved in all known IFN genes and in the HuIFN-β gene. Zinn *et al.* (1983) pointed out a large region of twofold rotational symmetry at nucleotides −201 to −164 in the HuIFN-β gene, and noted that the critical cutoff region for a likely inducer-responsive control region, which they localized to positions −74 to −76 in the interferon-β gene, contains the sequence AGA, which is conserved at approximately the same position in nine cloned HuIFN-α genes (Ragg and Weissmann 1983). Several other suggestive sequence comparisons can be made, but we do not yet have a clear understanding of their significance.

Equally suggestive sequence comparisons can be made between mouse and human interferon genes (only some of the former have been sequenced). Such comparisons are at least as important as those among human interferon genes, since in the heterologous expression experiments noted above, the exogenous human genes presumably responded to the same mediators of the induction signal that the host's genes normally respond to. Not only do both host and exogenous genes respond to the same induction signal, but as noted before in the case of HuIFN-α_1, the time course of appearance and disappearance of the HuIFN-α_1 RNA in mouse cells is similar to that of the mouse IFN mRNA in the same cells (Mantei and Weissmann 1982), suggesting not only a common induction mechanism but also a common shut-off mechanism.

Sequence comparisons of HuIFN-α_1 and MuIFN-α_1 show a high degree of homology in the approximately 150bp preceding the likely transcription initiation site. There is also considerable homology between HuIFNα_1, MuIFNα_1 and MuIFN-α_2 in their 5′ untranslated transcription sequences (Shaw *et al.* 1983). Additional sequencing studies of mouse α-interferons now under way (Kelley and Pitha, unpublished; Kelley *et al.* 1983) show marked 5′ flanking region homology between another mouse α-interferon and HuIFN-α_1, MuIFN-α_1, and -α_2. Highly conserved sequences are found in the −70 to −120 region, with a sharp decline in homology in sequences upstream of appriximately −140. The positioning of a highly conserved region (−70 to −120) within the 117bp 5′ region found to be required for induction of HuIFN-α_1 (Ragg and Weissmann 1983) strongly suggests that this conserved region plays a regulatory role.

2.6. CONCLUSION

Studies of interferon gene expression in homologous and heterologous systems have helped to consolidate our understanding of how interferon and other eukaryotic genes are regulated. It appears that the inducibility of interferon genes is dependent on a proximal 5′ flanking region, a finding analogous to that in several other inducible genes now being studied. Expression is also, however, probably modulated by other adjacent DNA sequences, and at the RNA level by selective degradation of interferon transcripts after induction, and perhaps by processing of initial long transcripts.

Numerous questions remain to be answered, however. Thus we must look forward to the isolation of intracellular factors that mediate the induction signal and control the regulatory sequences in interferon genes, perhaps by binding to those sequences. Further delineation of regulatory sequences is necessary, as well as an understanding of the functional significance of their markedly homologous regions. The interferon shut-off mechanism should be further investigated, as well as the molecular basis of cell-type specific expression.

The cloning of interferon genes provided the early means for studying control of interferon gene expression, and served as a prototype for similar studies of other eukaryotic genes. A number of new experimental methods, such as expressing heterologous genes carried on stable, episomal BPV-derived vectors and co-amplification of heterologous genes with a selectable marker like DHFR, will facilitate further advances toward understanding mechanisms of regulation in interferon and other eukaryotic genes.

Acknowledgements

The authors are indebted to N. Babu K. Raj and Kevin A. Kelley for thoughtful discussions, and to Barbara L. Schneider for indispensable skill in preparing the manuscript.

2.6. REFERENCES

Adolf, G. R. and Swetly, P. (1981). Effects of butyrate in induction and action of interferon. *Biochem. Biophys. Res. Commun.* **103,** 806–12.

—— and —— (1982). Interferon production in human hematopoietic cell lines: response to chemicals and characterization of interferons. *J. Interferon Res.* **2,** 261–70.

Baron, S., Dianzani, F., and Stanton, G. J. (1982). General consideration of the interferon system. *Tex. Rep. Biol. Med.* **41,** 1–13.

Brinster, R. L., Chen, H. Y., Warren, R., Sarthy, A., and Palmiter, R. D. (1982). Regulation of metallothionein–thymidine kinase fusion plasmids injected into mouse eggs. *Nature* **296,** 39–42.

Burke, D. C. (1982). The mechanism of interferon production. *Phil. Trans. R. Soc. Lond.* **B229,** 51–7.

Canaani, D. and Berg, P. (1982). Regulated expression of human inteferon-β_1 gene after transduction into cultured mouse and rabbit cells. *Proc. Nat. Acad. Sci. USA* **79,** 5166–70.

Cavalieri, R. L., Havell, E. A., Vilcek, J., and Pestka, S. (1977). Induction and decay of human fibroblast interferon mRNA. *Proc. Nat. Acad. Sci. USA* **74,** 4415–19.

Content J., De Wit, L., Tavernier, J., and Fiers, W. (1983). Human fibroblast interferon RNA transcripts of different sizes in poly(I).poly(C) induced cells. *Nuc. Acids Res.* **11,** 2627–38.

Dean, D. C., Knoll, B. J., Riser, M. E., and O'Malley, B. W. (1983). A 5'-flanking sequence essential for progesterone regulation of an ovalbumin fusion gene. *Nature* **305,** 551–4.

De Maeyer-Guignard, J., De Maeyer, E., and Montagnier, L. (1972). Interferon messenger RNA: translation in heterologous cells. *Proc. Nat. Acad. Sci. USA* **69,** 1203–7.

Derynck, R., Content, J., De Clercq, E., Volckaert, G., Tavernier, J., Devos, R., and Fiers, W. (1980). Isolation and structure of a human fibroblast interferon gene. *Nature* **285,** 542–7.

Fiers, W., Remaut, E., Devos, R., Cheroutre, H., Contreras, R., Gheysen, D., Degrave, W., Stanssens, P., Tavernier, J., Taya, Y., and Content, J. (1982). The human fibroblast and human immune interferon genes and their expression in homologous and heterologous cells. *Phil. Trans. R. Soc. Lond.* **B299,** 29–38.

Goeddel, D. V., Leung, D. W., Dull, T. J., Gross, M., Lawn, R. M., McCandliss, R., Seeburg, P. H., Ullrich, A., Yelverton, E., and Grey, P. W. (1981). The structure of eight distinct cloned human leukocyte interferon cDNAs. *Nature* **290,** 20–6.

Gross, G., Mayr, U., and Collins, J. (1981). New poly I-C inducible transcribed regions are linked to the human IFN-β gene in a genomic clone. In *The biology of the interferon system* (ed. E. De Maeyer, G. Gallaso, and H. Schellekens) pp. 85–90. Elsevier, Amsterdam.

Gurdon, J. B., Lane, C. D., Woodland, H. R., and Marbaix, G. (1971). Use of frog eggs and oocytes for the study of messenger RNA and its translation in living cells. *Nature* **233,** 177–82.

Hauser, H., Gross, G., Bruns, W., Hochkeppel, H.-K., Mayr, U., and Collins, J. (1982). Inducibility of human β-interferon gene in mouse L-cell clones. *Nature* **297,** 650–4.

Havell, E. A., Yip, Y. K., and Vilcek, J. (1977). Characteristics of human lymphoblastoid (Namalva) interferon. *J. Gen. Virol.* **38,** 51–9.
Hayes, T. G., Yip. Y. K., and Vilcek, J. (1979). Le interferon production by human fibroblasts. *Virology* **98,** 351-63.
Higashi, Y., Sokawa, Y., Watanabe, Y., Kawade, Y., Ohno, S., Takaoka, C., and Taniguchi, T. (1983). Structure and expression of a cloned cDNA for mouse interferon-β. *J. Biol. Chem.* **258,** 9522–29.
Karin, M., Haslinger, A., Holtgreve, H., Richards, R. I., Krauter, P., Westphal, H. M., and Beato, M. (1984). Characterization of DNA sequences through which cadmium and glucocorticoid hormones induce human metallothionein-II_A gene. *Nature* **308,** 513–19.
Kelley, K. A., Kozak, C. A., Dandoy, F., Sor, F., Skup, D., Windass, J. D., De Maeyer-Guignard, J., Pitha, P. M., and De Maeyer, E. (1983). Mapping of murine interferon α-genes to chromosome 4. *Gene* **25,** 181–8.
Kessel, M. and Khoury, G. (1983). Induction of cloned genes after transfer into eukaryotic cells. In *Gene amplification and analysis* (ed. T. S. Papas, M. Rosenberg, and J. G. Chirikjian) Vol. 3, pp. 233–60. Elsevier North Holland, Amsterdam.
McCormick, F., Trahey, M., Innis, M., Dieckmann, B., and Ringold, G. (1984). Inducible Expression of amplified human beta interferon genes in CHO cells. *Mol. Cell. Biol.* **4,** 166–72.
Mantei, N. and Weissmann, C. (1982). Controlled transcription of a human α-interferon gene introduced into mouse L cells. *Nature* **297,** 128–32.
Marcus, P. I. and Sekellick, M. J. (1977). Defective interfering particles with covalently lined [±] RNA induced interferon. *Nature* **266,** 815–19.
—— and —— (1980). Interferon induction by viruses. III. Vesicular stomatitis virus: interferon-inducing particle activity requires partial transcription of Gene N. *J. Gen. Virol.* **47,** 89–96.
Maroteaux, L., Kahana, C., Mory, Y., Groner, Y., and Revel, M. (1983). Sequences involved in the regulated expression of the human interferon-β_1 gene in recombinant SV40 DNA vectors replicating in monkey cells. *EMBO J.* **2,** 325–32.
May, L. T., Sehgal, P. B., La Forge, K. S., and Inouye, M. (1983). Expression of native α- and β-interferon genes in human cells. *Virology* **129,** 116–26.
Mitrani-Rosenbaum, S., Maroteaux, L., Mory, Y., Revel, M., and Howley, P. M. (1983). Inducible expression of the human interferon β_1 gene linked to a bovine papilloma virus DNA vector and maintained extrachromosomally in mouse cells. *Mol. Cell. Biol.* **3,** 233–40.
Mulvihill, E. R., Le Pennec, J.-P., and Chambon, P. (1982). Chicken oviduct progesterone receptor: location of specific regions of high-affinity binding in cloned DNA fragments of hormone-responsive genes. *Cell* **28,** 621–32.
Nagata, S., Mantei, N., and Weissmann, C. (1980*a*). The structure of one of the eight or more distinct chromosomal genes for human interferon-α. *Nature* **287,** 401–8.
—— Taira, H., Hall, A., Johnsrud, L., Streuli, M., Ecsodi, J., Boll, W., Cantell, K., and Weissmann, C. (1980*b*). Synthesis in *E. coli* of a polypeptide with human leukocyte interferon activity. *Nature* **284,** 316–20.
Ohno, S. and Taniguchi, T. (1982). Inducer-responsive expression of the cloned human interferon-β_1 gene introduced into cultured mouse cells. *Nuc. Acids Res.* **10,** 967–77.
—— and —— (1983). The 5′-flanking sequence of human interferon-β_1 is responsible for viral induction of transcription. *Nuc. Acids Res.* **11,** 5403–12.
Patient, R. (1984). DNA hybridization – beware. *Nature* **308,** 15–16.
Pelham, H. R. B. (1982). A regulatory upstream promoter element in the drosophila Hsp 70 heat-shock gene. *Cell* **30,** 517-28.
Pitha, P. M. (1981). Interferon induction with insolubilized polynucleotides and their

preparation. In *Methods in Enzymology* (ed. S. Petska) Vol. 78, pp. 236–42. Academic Press, New York.

—— Ciufo, D. M., Kellum, M., Raj, N. B. K., Reyes, G. R., and Hayward, G. S. (1982). Induction of human β-interferon synthesis with poly (rI.rC) in mouse cells transfected with cloned cDNA plasmids. *Proc. Nat. Acad. Sci. USA* **79,** 4337–41.

Ragg, H. and Weissmann, C. (1983). Not more than 117 base pairs of 5′-flanking sequence are required for inducible expression of a human IFN-α gene. *Nature* **303,** 439–42.

Raj, N. B. K. and Pitha, P. M. (1977). Relationship between interferon production and interferon messenger RNA synthesis in human fibroblasts. *Proc. Nat. Acad. Sci. USA* **74,** 1483–7.

—— and —— (1980). Synthesis of new proteins associated with the induction of interferon in human fibroblast cells. *Proc. Nat. Acad. Sci. USA* **77,** 4918–22.

—— and —— (1981). Analysis of interferon messenger RNA in human fibroblast cells induced to produce interferon. *Proc. Nat. Acad. Sci. USA* **78,** 7426–30.

—— and —— (1983). Two levels of regulation of β-interferon gene expression in human cells. *Proc. Nat. Acad. Sci. USA* **80,** 3923–27.

—— Kelley, K. A., Kellum, M., and Pitha, P. M. (1983). Regulation of interferon genes expression in Namalva cells induced to synthesize interferons. In *The biology of the interferon system* (eds. E. De Maeyer and H. Schellekens) pp. 57–62. Elsevier Amsterdam.

Reynolds, F. H. Jr., and Pitha, P. M. (1974). The induction of interferon and its messenger RNA in human fibroblasts. *Biochem. Biophys. Res. Commun.* **59,** 1023–29.

—— Premkumar, E., and Pitha, P. M. (1975). Interferon activity produced by translation of human interferon messenger RNA in cell-free ribosomal systems and in *Xenopus* oocytes. *Proc. Nat. Acad. Sci. USA* **72,** 4881–5.

Sehgal, P. B., Leyles, D. S., and Tamm, I. (1978). Superinduction of human fibroblast interferon production: further evidence for increased stability of interferon mRNA. *Virology* **89,** 186–98.

—— May, L. T., La Forge, K. S., and Inouye, M. (1982). Unusually long mRNA species coding for human α- and β-interferons. *Proc. Nat. Acad. Sci. USA* **79,** 6932–6.

Shaw, G. D., Boll, W., Taira, M., Montei, N., Lengyel, P., and Weissmann, C. (1983). Structure and expression of cloned murine IFN-α genes. *Nuc. Acids Res.* **11,** 555–73.

Shows, T. B., Sakaguchi, A. Y., Naylor, S. L., Goeddel, D. V., and Lawn, R. M. (1982). Clustering of leukocyte and fibroblast interferon genes on human chromosome 9. *Science* **218,** 373–4.

Shuttleworth, J., Morser, J., and Burke, D. C. (1982). Control of interferon mRNA levels and interferon yields in butyrate and 5′ bromodeoxyuridine treated Namalva cells. *J. Gen. Virol.* **58,** 25–35.

—— —— and —— (1983). Expression of interferon-α and interferon-β genes in human lymphoblastoid (Namalva) cells. *Eur. J. Biochem.* **133,** 399–404.

Stewart, E. W. II, Blalock, J. E., Burke, D. C., Chany, C., Dunnick, J. K., Falcoff, E., Friedman, R. M., Galasso, G. J., Joklik, W. K., Vilcek, J. T., Younger, J. S., and Zoon, K. C. (1980). Interferon nomenclature. *J. Immunol.* **125,** 2353.

Stewart, W. E. III (1979). *The interferon system.* Springer, Vienna.

Tan, Y. H., Armstrong, J. A., Ke, Y. H., and Ho, M. (1970). Regulation of cellular interferon production; enhancement by antimetabolites. *Proc. Nat. Acad. Sci. USA* **67,** 464–71.

Taniguchi, T., Guarente, L., Roberts, T. M., Kimelman, D., Douhan III, J., and Ptashner, M. (1980). Expression of the human fibroblast interferon gene in *E. coli.*

Proc. Nat. Acad. Sci. USA **77,** 5230–3.

Tavenier, J., Gheysen, D., Duerinck, F., Van der Heyden, J., and Fiers, W. (1983). Deletion mapping of the inducible promoter of human IFN-β gene. *Nature* 301, 634–6.

Vilcek, J., Havell, E. A., and Kohase, M. (1976). Superinduction of interferon with metabolic inhibitors: possible mechanisms and practical applications. *J. Infect. Dis.* **133,** A22-9.

Weidle, U. and Weissmann, C. (1983). The 5′-flanking region of a human IFN-α gene mediates viral induction of transcription. *Nature* **303,** 442–6.

Weissman, C., Nagata, S., Boll, W., Fountoulakis, M., Fujisawa, A., Fujisawa, J.-I., Haynes, J., Henco, K., Mantei, N., Ragg, H., Schein, C., Schmid, J., Shaw, G., Streuli, M., Taira, H., Todokoro, K., and Weidle, U. (1982). Structure and expression of human alpha-interferon genes. In *Interferons* (ed. T. C. Merigan and R. M. Friedman) pp. 295–326. Academic Press, London.

Wilson, V., Jeffreys, A. J., Barrie, P. A., Boseley, P. G., Slocombe, P. M., Easton, A., and Burke, D. C. (1983). A comparison of vertebrate interferon gene families detected by hybridization with human interferon DNA. *J. Mol. Biol.* **166,** 457–75.

Zinn, K., Mellon, P., Ptashne, M., and Maniatis, T. (1982). Regulated expression of an extrachromosomal human β-interferon gene in mouse cells. *Proc. Nat. Acad. Sci. USA* **79,** 4897–901.

Zinn, K., Di Maio, D., and Maniatis, T. (1983). Identification of two distinct regulatory regions adjacent to the human β-interferon gene. *Cell* **34,** 865–79.

3 Interferon and viruses: *in vitro* studies

B. R. G. Williams and E. N. Fish

3.1. INTRODUCTION

Interferons were originally characterized by their ability to protect cells from virus infection. They are now defined and standardized according to their inhibitory effect on the replication of specified viruses in selected cell types. Since interferons also affect a diverse number of biochemical responses in cells, it is perhaps not surprising that there is no single mechanism of replication inhibition to which different viruses are equally susceptible. The degree of inhibition observed is dependent both on cell type and virus group and is limited by the host range specificity of a given interferon. Interferons also show varying degrees of effectiveness on different types of cell cultures derived from the same species. These limitations in activity are now thought to result in part from subtle changes in the interaction of interferon with its cell surface receptor.

It has also become apparent that cells respond to interferon exposure by initiating a series of biochemical changes which eventually result in the development of an antiviral state. Not only is the activation of these biochemical changes or pathways different in different cell types but viruses are variously susceptible to their action. For example, encephalomyocarditis virus (EMC) and vesicular stomatis virus (VSV) are very sensitive to the antiviral activity of interferons and are widely used to titrate and standardize interferon activity. However, it has become clear that the growth of these two viruses is susceptible to inhibition by different biochemical pathways initiated by exposure of cells to interferon.

The three major types of interferon (alpha, beta and gamma) all induce the same overall biochemical changes in a given cell type and, as a consequence, have broadly the same spectrum of antiviral activities. The major well-characterized difference in the biochemical activities of the three groups is the induction of both the class I and class II major histocompatibility antigens by IFN-γ but not by INFS-α or -β. However, while this may have profound consequences for the immunoregulatory activities of different interferons, the effect on antiviral activities is less certain. If significant changes in membrane topography accompany the enhanced expression of cell surface antigens in response to interferon, then it might be expected that alterations in virus attachment and penetration into cells might occur. This has recently been demonstrated *in vitro* in interferon-treated VSV-infected cells (see below).

As indicated at the outset, a wide range of viruses are susceptible to interferon-induced inhibition. Although the systems used to measure interferon-induced viral inhibition (plaque reduction techniques, single cycle yield reduction, cytopathic effect inhibition) may differ, the relative sensitivities of different viruses to a particular interferon species are unaltered. Their respective sensitivities, however, vary with the viral type, the host cell, the type of interferon, and the viral infective dose. In order to understand the molecular basis of the antiviral action of interferon it is important, therefore, to define the contributions made by each component in the system.

3.2. COMPARATIVE SENSITIVITIES OF VIRUSES TO INTERFERONS

Early comparative studies involving a broad spectrum of viruses, both DNA and RNA containing, have identified those viruses that are most susceptible to interferon-induced inhibition (see Stewart 1979). Although the majority of comparative studies to date have involved the use of impure interferon preparations that were derived from a variety of sources, the relative sensitivities of different viruses to interferon-induced inhibition remain unchanged; where purer preparations of interferons are used the extent of inhibition improves. A number of DNA viruses, including adenoviruses and some members of the herpesvirus group, are minimally susceptible to interferon activity *in vitro* (Gallagher and Khoobyarin 1969; Glasgow, Hanshaw, Merigan, and Petralli 1967). The mechanisms for such 'interferon-resistance' are at present unclear, but may be related to the drastic inhibition of host cell metabolism elicited by these viruses. Experiments designed to examine the interferon sensitivities of representative adenovirus types by means of plaque reduction assays, demonstrated that similar patterns of interferon sensitivity were noted for the adenovirus types 2, 7, and 12 (Gallagher and Khoobyarin 1969). In yield reduction assays, rhinovirus strains 2 and 4 and influenza virus strains A2 and B have been shown to be comparable in their interferon sensitivity to VSV, a virus known to be interferon-sensitive (Merigan, Reed, Hall, and Tyrrell 1973). The implications are that within a particular virus group the relative sensitivity to interferon-induced viral inhibition is constant. An exception to this general rule however, is seen with the herpesviruses. Protection from viral infection is far more pronounced with herpesvirus type I (HSV-1) than with herpesvirus type 2 (HSV-2) (Stanwick, Schinazi, Campbell, and Nahmias 1981; see Table 3.1).

The differential sensitivities exhibited by viruses to a particular interferon preparation reflect the magnitude and extent to which an interferon induces

Table 3.1. *Comparative sensitivities of viruses to different IFNs*

IFN	Cell type	Virus	m.o.i.	ID_{50} i.u. ml^{-1} [a]	Reference
IFN-αLy	HeLa	Polio	3	3.8	Munoz and Carrasco 1981
IFN-αLy	HeLa	EMC	0.1	3	Munoz and Carrasco 1984
IFN-αLy	HeLa	VSV	0.1	5.6	
IFN-αLy	HeLa	HSV-1	0.01	40	
IFN-αLy	Daudi	VSV	200	280	Yonehara, Ishii, and Yonehara-Takahashii 1983
IFN-αLy	T98G	HSV-1	0.1	5	Fish *et al.* 1983
IFN-αLe	HeLa	EMC	0.1	3	Munoz and Carrasco 1984
IFN-αLe	HeLa	VSV	0.1	9	
IFN-αLe	HeLa	HSV-1	0.01	5	
IFN-β	HeLa	EMC	0.1	3	
IFN-β	HeLa	VSV	0.1	3	
IFN-β	HeLa	HSV-1	0.01	7.3	
IFN-β	HeLa	VSV	10	23	Wallach 1983
IFN-β	Vero	VSV	200	1	Stanwick *et al.* 1981
IFN-β	Vero	HSV-1	200	53	
IFN-β	Vero	HSV-2	200	1500	
IFN-γ	HeLa	EMC	0.1	3	Munoz and Carrasco 1984
IFN-γ	HeLa	VSV	0.1	11	
IFN-γ	HeLa	HSV-1	0.01	24	

[a]IFN concentration, i.u. ml^{-1}, causing 50 per cent viral inhibition.

an inhibitory effect at a particular stage along a pathway in the replicative cycle of the virus. For different viruses these sites of inhibitory action will be different, and the ability of an interferon to arrest viral growth will vary accordingly. Even though the replicative cycles of HSV-1 and HSV-2 show only minor differences, these are clearly significant enough to lead to differential sensitivity to interferon.

3.3. HOST CELL SPECIFIED INHIBITION

Studies on the mechanism of interferon action by Samuel and Knutson (1982*a,b*) have shown that different subspecies of biologically active interferon induce equivalent antiviral activities and biochemical changes in mouse L929 cells and that the kinetics of induction and decay of the antiviral state are dependent on the cell type. Reports comparing the interferon sensitivities of different viruses on different cell types (see Stewart 1979) demonstrate that the extent of interferon-induced resistance is, indeed, determined by the treated cell rather than by the interferon, i.e. the type of resistance induced is characterized by the cell type. An example of this cell-specified effect is seen when comparing the interferon-induced inhibition of VSV in bovine and porcine cells: VSV is approximately 10^3-fold more sensitive than bovine enterovirus to interferon-induced inhibition in bovine cells, yet is approximately 20-fold less sensitive than this virus in porcine cells (Ahl and Rump 1976). Raji and Daudi, two Burkitt's lymphoma cell lines that are equally susceptible to superinfection with Epstein-Barr virus, respond differently after exposure to interferon. In Daudi cells, less than 10 i.u. of human leukocyte interferon (IFN-αLe) is sufficient to prevent expression of virally induced early antigen in 50 per cent of the infected cells (Adams, Lidin, Strander, and Cantell 1975). In contrast, at least a 100-fold higher concentration of exogenous interferon is needed to similarly inhibit viral gene expression in Raji cells (Lidin and Lamon 1982).

3.4. DIFFERENTIAL ACTIVITIES OF DIFFERENT INTERFERONS

Another component to be considered when examining the relative sensitivities of different viruses to interferons is the extent of inhibition induced by a particular molecular species of interferon. Although natural IFN-α preparations are mixtures of various subtypes, it appears that different cells and induction procedures result in different ratios of the various subtypes (Allen and Fantes 1980; Rubinstein, Levy, Moschera, Lai, Hershberg, Bartlett, and Pestka 1981; Evinger, Rubinstein, and Pestka 1981). In leukocyte interferon, IFN-αA and IFN-αD (as defined by Goeddel, Leung, Dull, Gross, Lawn, McCandliss, Seebury, Ullrich, Yelverton, and Gray 1981 and Weck, Apperson, May and Stebbing 1981a) appear to be the predominant species, comprising approximately 80 per cent of the total species present (Goeddel *et al.* 1981; Levy, Shirely, Rubinstein, Valhe, and Pestka 1980). Lymphoblastoid interferon (IFN-α-Ly) contains predominantly α interferons, with approximately 15 per cent IFN-β (Havell, Yip, and Vilcek 1977; Dalton and Paucker 1979). The major species in IFN-αLy appear to correspond to IFN-αB, IFN-αF or IFN-αI (Zoon, Smith, Bridgen, Anfisen, Hunkapiller, and Hood 1980). As shown in Table 3.1, the extent of viral inhibition in a given cell type for a given virus challenge is determined by the interferon type. Whereas IFN-αLy, IFN-αLE, IFN-β, or IFN-γ treatment of Hela cells leads to comparable inhibition with respect to EMC virus infection, the same interferons are differentially active with respect to VSV or herpes simplex virus type I (HSV-I) infections (Munoz and Carrasco 1984).

Recent studies with pure preparations of both natural and recombinant DNA-derived interferons have revealed that variations in the magnitude of interferon-mediated responses are exerted by different subtypes of interferons (Table 3.2) as well as by the different classes of interferons. Individual IFN-αLe subtypes show distinct antiviral effects against different viruses in a

Table 3.2. *Antiviral activities of recombinant DNA-derived IFN-αs*

IFN	Cell type	Virus	m.o.i	ID_{50} i.u. ml^{-1} [a]	Reference
IFN-αA	K562	EMC	0.1	175	
IFN-αC	K562	EMC	0.1	500	
IFN-αD	K562	EMC	0.1	200	
IFN-αAD	K562	EMC	0.1	70	Fish *et al.* 1983
IFN-αA	K562	HSV-2	0.1	145	
IFN-αD	K562	HSV-2	0.1	250	
IFN-αAD (Bgl)	K562	HSV-2	0.1	37	
IFN-αAD (Bgl)	T98G	EMC	0.1	600	
IFN-αAD (Bgl)	T98G	HSV-2	0.1	500	
IFN-αD	T98G	HSV-2	0.1	430	
IFN-αD	T98G	EMC	0.1	250	
IFN-αAD (Bgl)	WISH	VSV	0.01	1	
IFN-αAD (Bgl)	HeLa	VSV	0.01	1	
IFN-αAD (Pvu)	WISH	VSV	0.01	1	
IFN-αDA (Bgl)	WISH	VSV	0.01	1	
IFN-αDA (Bgl)	HeLa	VSV	0.01	1.6	Weck, *et al.* 1981*b*
IFN-αDA (Pvu)	WISH	VSV	0.01	1	
IFN-αDA (Pvu)	HeLa	VSV	0.01	0.13	
IFN-αD	WISH	VSV	0.01	1	
IFN-αD	HeLa	VSV	0.01	32	
IFN-αA	WISH	VSV	0.01	1	
IFN-αA	HeLa	VSV	0.01	100	
IFN-αA	Hu Aminon U	Reovirus	10	300	Samuel and Knutson 1981

[a]IFN concentration, i.u. ml,$^{-1}$ causing 50 per cent viral inhibition.

variety of *in vitro* cell systems (Masucci, Szigeti, Klein, Gruest, Taira, Hall, Nagata, and Weissmann 1980; Weck *et al.* 1981*a*; Fish, Banerjee and Stebbing 1983). In addition, hybrid interferons derived from these subtypes also show distinct activities. Whereas mixtures of the two subtypes IFN-αA and IFN-αD show additive effects, the IFN-αAD hybrids derived from these parental subtypes have greater activities and the reverse hybrids, IFN-αDA, generally much lower activities (Weck, Apperson, Stebbing, Gray, Leung, Sheppard, and Goeddel 1981*b*; Fish, *et al.* 1983). To some extent these differential efficacies exhibited by the different subtypes and hybrids may be explained by differential cell surface receptor binding interactions. Variations in interferon sensitivities exhibited by different cell types may be related to the number of interferon cell surface receptors and/or the relative binding affinities of the different interferon subtypes and hybrids to those receptors (Branca and Baglioni 1981; Hannigan, Gewert, Fish, Read, and Williams 1983). However, the level of functional activity of an interferon cannot be entirely explained in terms of cell surface receptor binding activities since the subtypes IFN-αA and IFN-αD that apparently demonstrate different affinities for IFN-α cell surface receptors in competition binding studies on Daudi cells, show comparable biological activities *in vitro* (Hannigan *et al.* 1983).

Subtle structural differences among the different recombinant interferons could account for the differences in antiviral activities observed *in vitro*. The various cloned leukocyte IFN subtypes have been classified into two major groups and an intermediate group, according to alternative amino acid residues to be found at defined positions, namely 14, 16, 71, 78, 79, 83, 154 and 160 (Weissmann 1981). IFN-αA and -D fall into group I whereas IFN-αB, -C and -F fall into group II, or the intermediate group (Weissmann 1981). Interestingly, these interferons may also be assigned to the same two groups on the basis of their antiviral activities: the antiviral activities of IFN-αA and IFN-αD are comparable and greater than IFN-αB, -C or -F (Fish *et al.* 1983). Thus, the structural differences cited may be the basis for the differences in observed antiviral activity. The generation of hybrid IFN-αs has resulted in some preliminary structure/function studies, yet no definitive data relating particular structural domains with antiviral activity is available. The differences in antiviral efficacy between IFN-αDA (Bgl II) and either of the parental subtypes or the reverse hybrid supports results indicating that neither the N- or C-terminal portions of the interferon molecule alone determine cell receptor binding (Streuli, Hall, Boll, Stewart, Nagata, and Weissmann 1981; Hannigan *et al.* 1983).

3.5. SYNERGISTIC ACTIVITIES OF INTERFERONS *IN VITRO*

Interferon-induced antiviral inhibition requires the interaction of interferon with specific cellular receptors. Receptor binding studies have identified the existence of at least two functional interferon receptors on mouse (Aguet, Belardelli, Blanchard, Marcucci, and Gresser 1982; Hovanessian, La Bonnardiere, and Falcoff 1980) and human (Anderson, Yip, and Vilcek 1982; Branca and Baglioni 1981; Hannigan, Fish, and Williams 1984) cells. The indications are that IFN-α or IFN-β bind to one type of receptor, whereas IFN-γ binds to another. The existence of two functionally distinct interferon receptors supports the hypothesis that different mechanisms of inhibitory action associated with different interferons are initiated as early as the interaction of the interferon molecule with its cell surface receptor. Substances that have identical modes of action generally interact in an additive manner. A synergistic response resulting from treatment with two agents in combination implies that these two agents exert their activities through distinct pathways. The demonstration of antiviral synergistic responses with combinations of type I and type II interferons, and not combinations of type I interferons, further supports the concept of differences in the pathways induced by these interferons. For example, partially purified preparations of mouse interferon-γ and mouse interferon-α/β potentiate each other with respect to the inhibition of mengovirus (Fleischmann, Georgiades, Osborne, and Johnson 1979), EMCV and HSV-1 (Zerial, Hovanessian, Stefanos, Huygen, Weiner, and Falcoff 1982) replication in murine cells. The treatment of human melanoma or embryonic lung carcinoma cells with recombinant HuIFN-γ in combination with either recombinant HuIFN-α or HuIFN-β results in potentiation of interferon-induced inhibition of HSV-1 replication (Czarniecki, Fennie, Powers, and Estell 1984). Preparations of recombinant human interferon-α and interferon-γ also potentiate each other with respect to the inhibition of VSV in human amnion U cells (Samuel and Knutson 1983). Although this synergism observed for combinations of type I and type II interferons may reflect the independent interactions of these interferons with their distinct cell surface receptors, recent studies have shown that the binding of human interferon-γ to its distinct cell surface receptor can modify expression of the separate interferon-α receptor, resulting in an inhibition of interferon-α induced antiviral activity (Hannigan *et al.* 1984).

The differential effects of type I and type II interferons on the induction of both the synthesis of 2-5A synthetase and HLA antigens in human lymphoblastoid and fibroblastoid cells has been demonstrated; 2-5A synthetase was induced poorly by IFN-γ as compared to IFN-α or -β (Wallach, Fellous, and Revel 1982). This differential induction of 2-5A synthetase is also seen in human peripheral blood lymphocytes (S. E. Read and B. R. G. Williams, unpublished observations). Therefore, while the initial interaction of the interferon with its specific cell surface receptor is characterized by the host cell, it is clear that resultant antiviral activity is determined by the nature and extent of biochemical pathways induced.

3.6. INTERFERON IN COMBINATION WITH ANTIVIRAL AGENTS *IN VITRO*

Antiviral synergism has also been demonstrated when interferons are used in combination with other antiviral agents *in vitro*. These studies have centred almost exclusively on the herpesviruses and those antiviral compounds that have been shown to have limited efficacy against them; namely 9-β-D arabinofuranoxyladenine (ara A), 1-β-D-arabinofuranoxylcytosine (ara C), (E)-5-(2-bromovinyl-2′-deoxyuridine (BVDU), acyclovir (ACV) and, more recently, the acyclic guanine nucleoside 9-[(1,3-dihydroxy-2-propoxy) methyl] guanine (DHPG). It is thought that these antiviral agents exert their effect by a mechanism involving phosphorylation of the agent by a thymidine kinase, and subsequent inhibition of viral transcription by directly or indirectly inhibiting the viral DNA polymerase. The relative accumulation of these different phosphorylated products in infected cells appears to correlate with their relative efficacies as antiviral agents.

Since interferons have not been shown to significantly inhibit viral DNA synthesis one might anticipate a potential for synergism when they are used in combination with these antiviral agents. This antiviral synergism could result from the inhibition of different stages of viral replication by the different drugs, resulting in a potentiated inhibition of viral replication. This is indeed the case. Marked synergistic inhibition of HSV-2 has been obtained by *in vitro* combinations of DHPG or ACV with recombinant HuIFN-α, -β, or -γ (Eppstein and Marsh 1984). Significantly more potent anti-HSV-2 activity was obtained by the synergistic combination of interferon with DHPG than with ACV, presumably due to the more efficient accumulation of DHPG triphosphate in infected cells. Combinations of interferon-α with ACV (Levin and Leary 1981) and IFN-β with ACV (Stanwick *et al.* 1981) potentiate each other with respect to inhibition of HSV-1, HSV-2 and varicella zoster virus (VZV). Synergistic HSV inhibitory interactions have been reported with ara-A in combination with IFN-α (Lerner and Bailey 1974). Whereas one might anticipate that combination treatments involving interferons with ara A, ara C, BVDU or indeed ACV would without exception potentiate inhibition of viral replication, this is not always the case. The anti-HCMV effect of ACV combined with IFN-α (Levin and Leary 1981) or IFN-β (Spector, Tyndall, and Kelley 1982) has been demonstrated to be additive, and, depending on the HCMV strain, to be additive to synergistic. Plaque reduction assays indicated that the additive interaction of ACV and IFN-α could be changed to a synergistic interaction by increasing the concentration of ACV (Smith, Wigdahl, and Rapp 1983). The necessity to supplement cultures with additional ACV to demonstrate inhibition of HCMV plaque formation by ACV corroborates the finding that ACV exerts only a transient inhibition on the synthesis of 2 HCMV-specific late polypeptides (Mar, Patel, and Huang 1982). With ara-A or ara-C, increasing the concentrations of these nucleosides to test for a synergistic interaction with IFN-α is precluded, as cellular toxicity develops.

A major source of discrepancy for the demonstration of an additive or synergistic antiviral effect is the multiplicity of infection (m.o.i.) used in experiments. It is clear that synergism is only seen when the interferon/antiviral to m.o.i. ratio is high, and that the sensitivity of a given virus is critically dependent on this ratio. The higher this ratio, the better the protection of *cellular* mRNA translation achieved. For example, in interferon-treated cells infected at a high m.o.i. with EMC virus, shut-off of cellular protein synthesis occurs and a cytopathic effect develops with a timing at the same rate as in cells not treated with interferon. Eventually the EMC virus-infected cells die, even if interferon is continually added to the cultural medium (Munoz and Carrasco 1981). This occurs despite evidence that the interferon-induced 2-5A synthetase system is active under these circumstances (Williams, Golgher, Brown, Gilbert, and Kerr 1979*b*).

3.7. MECHANISMS OF ANTIVIRAL ACTIVITY *IN VITRO*

As described above, the resistance to virus infection induced by interferon is widely based and is usually a characteristic of the cells rather than of the interferon. This resistance may be manifested at any one or more of the following sites during virus replication.

1. Early events, including attachment, penetration, uncoating, and early transcription.
2. Translation.
3. Assembly, maturation, and release.

3.7.1. Interferon inhibition of early stages of virus growth

Until recently there was little compelling evidence that early stages in virus replication, i.e. before the expression of virus specific functions, were significantly affected by interferon. However, it has been known for some time that interferon treatment can drastically alter the structure and organization of the membrane and cytoskeleton of a variety of cells, and has an inhibitory effect on pinocytosis (Tamm, Wang, Landsberger, and Pfeffer 1982). Therefore, it was perhaps not surprising that a recent examination of VSV entry into interferon-treated cells detected a strong inhibition of receptor-mediated endocytosis (Whitaker-Dowling, Wilcox, Widnell, and Younger 1983). Interestingly, this effect does not always correlate with the development of an antiviral state. Some cell lines do not respond to interferon treatment with reduced pinocytosis yet do develop resistance to virus infection (Whitaker-Dowling *et al.* 1983). While the concentrations of interferon required to elicit the antipinocytic effect are relatively high compared to that required to induce 2-5A synthetase (see below), at low multiplicities of infection and under physiological conditions, a considerable inhibitory effect on virus growth does occur. Viruses that enter cells by direct fusion with the membrane (e.g. members of the paramyxovirus group) may not be subject to inhibition at this stage of their replicative cycle. However, beta-propiolactone-inactivated Sendai virus-mediated cell fusion is inhibited in virus-infected interferon-treated cells (Chatterjee, Cheung, and Hunter 1982). Nevertheless, it is likely that there exists a differential effect of interferon at this level on different viruses.

A well characterized early event inhibited by interferon treatment occurs in SV40 infected cells. SV40 RNA and DNA production early in infection can be markedly inhibited in cells previously exposed to interferon (Brennan and Stark 1983). This inhibition is significantly decreased with increasing multiplicities of infection. The inhibitory step is apparently directed at the onset of transcription from partially uncoated virions. This could either result from the failure to remove a virus component which prevents transcription or the presence of an interferon-induced transacting cellular repressor which specifically inhibits the onset of viral transcription. It appears therefore that the uncoating of SV40 DNA is inhibited in cells pretreated with interferon. It might also be expected that early transcription in other DNA tumour viruses could be a target for interferon action. Indeed, an early step in the replication of HSV-1 is susceptible to inhibition by interferon treatment (Panet and Falk 1983). The induction of both thymidine kinase (tk) and DNA polymerase activity can be inhibited, followed by a subsequent reduction in virus yield. This effect is dependent on multiplicity of infection and is not seen when an m.o.i. higher than 10 is used. The sensitive step appears to be the expression of immediate-early HSV-1 alpha genes. Interferon had no effect on the enzyme activity of a cloned tk gene stably integrated into cellular DNA and constitutively expressed.

The early events in retrovirus replication are also subject to inhibition by interferon. There appears to be an interferon-mediated block before integration and moderate effects on both virus uncoating and reverse transcriptase-directed DNA synthesis (for example see Avery, Norton, Jones, Burke, and Morris 1980 and Aboud, Shoor and Salzberg 1980). While these effects appear to be only transient, a more dramatic effect on the supercoiling of the proviral DNA is seen in normal rat kidney cells infected with Moloney murine sarcoma virus. Since supercoiling is a requirement for integration, inhibition of this step results in the accumulation and subsequent degradation of unintegrated viral DNA (Huleihel and Aboud 1983). The selection of cells and virus are likely to be important in demonstrating this early interferon effect. Studies in other systems have not detected an early inhibition of retrovirus replication (see Riggin and Pitha 1982).

This phenomenon or a similar inhibitory effect on DNA integration might also be responsible for the reported inhibition by interferon of transformation of mouse cells by plasmids containing the herpes virus tk gene or the dihydrofolate reductase gene (Dubois, Vignal, Le Cunff, and Chany 1983). Transport of DNA from the cytoplasm to the nucleus did not seem to be inhibited and there may have been an enhanced but transient expression of the tk gene in interferon-treated cells. This may parallel the anti-retrovirus effect, described above, with an accumulation of unintegrated but expressible DNA in the nucleus.

Several reports have described the inhibition of the appearance of transformation-related phenotypes in cells transformed by a variety of agents. This is probably due to the combination of interferon effects on membranes and cytoskeleton organization. One recent publication has reported a selective reduction in the synthesis of pp60src in Rous sarcoma virus-transformed cells treated with interferon (Lin, Garber, Wang, Caliguiri, Schellekens, Goldberg, and Tamm 1983). This correlates with reduced levels of pp60 src-associated tyrosine phosphokinase activity. This is an interesting observation because it suggests that interferon may be able to selectively inhibit the expression of an integrated retroviral gene. Confirmation of this hypothesis using specific hybridization probes for the pp60src mRNA is required before this can be accepted as a defined site of interferon action. It may be that an interferon-induced alteration in the usual sites of accumulation of pp60src in the cell membrane indirectly leads to a reduction in pp60src synthesis.

For other DNA and RNA viruses there is little evidence that primary transcription is significantly affected by interferon-induced inhibitory mechanisms. Furthermore, any analysis of such an effect must first exclude the involvement of the 2-5A synthetase system described below.

3.7.2. Interferon-mediated inhibition of translation

During the past several years, much attention has been focused on the inhibition of virus replication at the level of mRNA translation. There is now good evidence for the activation of at least two biochemical pathways in interferon-treated virus-infected cells that lead to inhibition of protein synthesis. Both these pathways are dependent on the presence of double-stranded RNA which may be produced either as a replicative intermediate or as a result of the transcriptional patterns of viruses. In the best studied pathway, the activation of a 2′, 5′-oligoadenylate synthetase (2-5A synthetase) by double stranded RNA leads to the formation from ATP of 2′, 5′-linked oligonucleotides of adenosine. These oligonucleotides activate an endogenous cellular endonuclease which degrades both messenger and ribosomal RNA. Another less well characterized pathway involves the phosphorylation of the eukaryotic initiation factor by a double stranded RNA-dependent protein kinase (see Williams 1983, and Johnstone and Torrence 1984 for recent reviews of these pathways). While these pathways certainly appear active in mediating the antiviral effects of interferon in some virus–cell interactions (discussed below), viruses which do not generate dsRNA may not be affected. The ability to accurately measure the activation of the 2-5A synthetase pathway in particular, in the same cells infected with different viruses, suggests that alternative mechanisms of action probably exist. Attention is now increasingly being given to other possible biochemical mechanisms of action. Some of the recent studies are discussed below and all serve to again draw attention to the variability seen with different virus–cell systems.

The interferon-induced protein kinase

The overall significance of protein kinase in the interferon-mediated inhibition of viral mRNA translation remains uncertain. Studies attempting to correlate the antiviral action of interferons with the presence or absence of protein kinase activity have proved contradictory. For example, correlations have been made between the induction of protein kinase activity by IFN-α_2 in human amnion cells and inhibition of VSV replication under conditions where reovirus was unaffected (Samuel and Knutson 1981). IFN-γ, on the other hand, did not induce the kinase but was effective in establishing an antiviral state against both VSV and reovirus. The antiviral activities of IFN-α_2 and IFN-γ were synergistic against VSV only but this did not correlate with an increase in protein kinase activity (Samuel and Knutson 1983). A lack of correlation between the induction of protein kinase activity and inhibition of VSV replication has been reported in other cell types by other laboratories (Hovanessian, Meurs, and Montagnier 1981; Sen and Herz 1983). Therefore, the role of this enzyme in the action of interferon against VSV is uncertain.

Reovirus infection of interferon-treated HeLa cells results in the elevation of kinase activity and the concomitant phosphorylation of initiation factor eIF2. Two HeLa lines showing different sensitivities to interferon-induced inhibition of reovirus growth had comparable levels of 2-5A synthetase activity, but the more sensitive line demonstrated 3–4 times more protein kinase activity (Nilsen, Maroney, and Baglioni 1982*a*). Clearly the relative importance of the protein kinase in mediating the antiviral effects of interferon is dependent on the virus–cell system being studied and perhaps on the type (alpha or gamma) of interferon being used. A preactivated protein kinase has been detected in interferon-treated EMC and reovirus virus-infected cells (Golgher, Williams, Gilbert, Brown, and Kerr 1980; Nilsen *et al.* 1982*a*; Gupta, Holmes, and Mehra 1982). In contrast, infection with mengovirus or VSV did not result in demonstrable activation of the kinase. It is difficult to reconcile the differences in kinase activation between mengovirus- and EMC-virus-infected cells and, in fact, the first report of enhanced interferon-dependent kinase activity described its detection in mengovirus-infected mouse L cells (Aujean, Sanceau, Falcoff, and Falcoff 1979). Once again differences in cell lines must be invoked to help explain this discrepancy.

An interesting recent development concerning the protein kinase is the description by Rice and Kerr (1984) of an inhibition of double-stranded RNA-dependent kinase activity in interferon-treated vaccinia virus-infected cells. This virus-mediated action makes it unlikely that the kinase plays a role in inhibition of virus growth under the conditions investigated. Although both the 2-5A synthetase and protein kinase enzymes require dsRNA for activation, the kinase requirements appear to be subtly different from the synthetase. Much lower concentrations of dsRNA will activate the kinase in cell-free systems, and low KCl and high Mg^{2+} also increase its activity. High concentrations of dsRNA, on the other hand, inhibit kinase activity but effectively activate the synthetase (Williams, Gilbert, and Kerr 1979*a*). Clearly there exists a mechanism for selectively activating these enzyme activities. The interactions of specific viruses with different interferon-treated cells resulting in production of varying concentrations of dsRNA or differing intracellular ionic conditions may ultimately determine which, if either, of the 2-5A synthetase or protein kinase pathways operate to inhibit virus growth.

The 2-5A synthetase pathway

There is now good evidence for the activation of the 2-5A synthetase system in some specific cases of interferon-treated virus infected cells. Both 2-5A itself and/or endonuclease activation have been detected in EMC virus-infected mouse L and human HeLa cells (Williams, Golgher, Brown, Gilbert, and Kerr 1979*b*; Silverman, Cayley, Knight, Gilbert, and Kerr 1982), mengovirus infected L cells (Vacquero, Sanceau, and Falcoff 1982), reovirus-infected HeLa cells (Nilsen, Maroney, and Baglioni 1982*b*) and vaccinia virus-infected L cells (Goswami and Sharma 1984). It appears reasonable to assume that the interferon-mediated inhibition of virus growth in these systems is through activation of the 2-5A synthetase system. However, once again there are exceptions. In SV40 virus-infected monkey CVI cells treated with interferon 8 h after infection, high levels of 2′-5′-oligoadenylates can be detected some 25 to 72 h later (Hersh, Brown, Roberts, Swryd, Kerr, and Stark 1984). These cells are not protected by this exposure to interferon and in accord with this only a small percentage of the oligomers found consisted of functional 2-5A. Furthermore, there was little evidence of 2-5A mediated endonuclease activity. The identity of the unusual 2′-5′-oligoadenylates present at such high levels in these cells remains unknown, but the authors speculate that perhaps an SV40 coded or induced protein may be able to circumvent the activation of a nuclease by conversion of 2-5A to inactive oligomers. Alternatively, or additionally, efficient compartmentalization of the parts of the viral replicative and the 2-5A synthetase pathways may result in avoidance of any interruption of mRNA activity. Further support for this comes from a recent report (Rice, Roberts, and Kerr 1984) describing the detection of synthesis of high concentrations of 2-5A in interferon-treated vaccinia virus infected cells. Although this 2-5A was active in cell-free systems, no degradation of RNA was apparent in the cells from which this material was derived. At first sight, these results appear to contradict the findings of Goswami and Sharma (1984). However, in the experiments described by Rice and co-workers no early inhibition of mRNA synthesis was seen. Therefore the effect of the interferon-induced 2-5A synthetase system on vaccinia virus replication may have two different outcomes even in two closely related cell lines. If the system is not activated early in infection then perhaps a separate compartmentalization process which is part of the virus replicative cycle, but may be subtly different in different cells, protects from further inhibition. There now exist a number of sensitive assays for various components of the 2-5A system (see Johnstone and Torrence 1984 for references). Furthermore, a molecular hybridization probe for 2-5A synthetase mRNA has now been described (Merlin, Chebath, Benech, Metz, and Revel 1983). When used in conjunction, these should be useful for determining activation and compartmentalization of the 2-5A system during virus infection. An earlier experiment performed by Nilsen and Baglioni (1979) demonstrated that single-stranded RNAs covalently joined to dsRNA segments are preferentially cleaved over identical RNA not bound to dsRNA. This lends credence to the idea that the 2-5A system could be effectively compartmentalized *in vivo* by the complexing of the enzymes involved (synthetase, endonuclease, and 2′-phosphodiesterase) to dsRNA.

Until recently, one of the puzzling findings related to the 2-5A system was the apparently normal levels of replication of some viruses in cells containing high basal levels of 2-5A synthetase. Subsequently it has been determined that a virus-directed inhibition of the 2-5A-dependent endonuclease occurs in these cells (Cayley, Knight, and Kerr 1982). Pretreatment of the cells with interferon can prevent this inhibition. The extent to which this occurs and whether it extends to viruses other than EMC has yet to be documented. Some viruses may be able to overcome the interferon effect on the endonuclease or this effect may be restricted to certain cell types. Part of the interferon effect on the nuclease may involve induction of its mRNA. Treatment of Daudi, HeLa or mouse L cells with interferon results in a two-fold increase in the levels of nuclease (Silverman, Krause, Jacobsen, Leisy, Barlow, and Friedman 1983). Ten- to twenty-fold increases have been found in another murine cell line (Jacobsen, Krause, Friedman, and Silverman 1983). Thus the relative importance of the 2-5A system in the antiviral activity of interferon may depend ultimately on the availability of activatible endonuclease.

The oligonucleotide 2-5A, when introduced directly into intact virus infected cells, can inhibit viral replication (reviewed by Johnstone and Torrence 1984). This effect of 2-5A is transient as the oligonucleotide is rapidly degraded by 2′-phosphodiesterase activity. To circumvent this, 2-5A analogues have been synthesized and tested as antiviral agents. Results show that at least some of these agents, while active as antiviral agents, may not activate the endonuclease. For example, it has been demonstrated that a xyloadenosine analogue of 2-5A inhibits the replication of herpesvirus types 1 and 2 Eppstein, Barnett, Marsh, Gosselin, and Imbach 1983; Eppstein, Marsh, and Schryver, 1984) through the slow release of xyloAMP which is probably phosphorylated and incorporated as a chain terminator into RNA. Similarly, a cordycepin analogue of 2-5A which can inhibit Epstein-Barr transformation of lymphocytes does not activate the endonuclease and presumably works through some alternative mechanism (Doetsch, Suhadolnik, Sawada, Mosca, Flick, Reinchenbach, Dang, Charubala, Pfleiderer, and Henderson 1981; Sawai, Imai, Lesiak, Johnston, and Torrence 1983). In the meantime, the use of 2-5A analogues as potential antiviral agents remains somewhat uncertain, although they do have potential to act as antagonists of the interferon-induced 2-5A system.

3.7.3. Interferon effects on virus maturation

Interferon has a significant effect on the latter stages of replication of retroviruses under conditions where there is no apparent effect on protein synthesis. The inhibitory effect of interferon either results in a reduction in the release of virus particles from the cell accompanied by an accumulation of budding virions at the cell membrane surface or the release of progeny virions with decreased infectivity. Recently it has become clear that this effect is not restricted to retroviruses but is probably common to all viruses that bud from the cell membrane. Although the yield of VSV particles from cells treated with low amounts of interferon is reduced by only 10-fold, the progeny VSV particles are reduced in infectivity by about 1000-fold (Maheshari, Banerjee, Waechter, Olden, and Friedman 1980). This appears to result from the specific inhibition of incorporation of the G and M proteins into the maturing VSV virions in interferon-treated cells (Jay, Dawood, and Friedman 1983). The reduction in glycosylation activity reported in interferon treated cells (Maheshwari *et al.* 1980) does not appear to be involved in this effect and the mechanism remains to be determined.

Vaccinia virus particles from interferon-treated infected cells also appear to be reduced in infectivity. There appears to be a reduction in both phosphorylation and glycosylation of particle proteins and this is accompanied by a decrease in adsorption, penetration, and uncoating when compared to particles from untreated cells (Esteban 1984). There is little doubt these observations will soon be extended to other viruses, but whether a common mechanism prevails is more uncertain. It is also apparent that even closely related cell lines may demonstrate significant variation in the degree of inhibition of this step in virus replication.

3.8. THE MX GENE

Alleles of a mouse gene, Mx, selectively influence the ability of interferon to protect against influenza virus infection. This can be demonstrated *in vitro* in peritoneal macrophages or mouse embryo fibroblasts and is perhaps the most striking example of a differential effect of interferon against different viruses (reviewed by Haller 1981). Cells derived from mice carrying the Mx gene can be protected against influenza virus infection by much lower concentrations of interferon than cells from non-Mx bearing strains. However, no difference is seen in the protection from other viruses. The kinetics of protection against influenza in Mx-bearing cells is markedly different from the antiviral state induced against VSV in these cells. This suggests that the two antiviral mechanisms are unrelated (Arnheiter and Haller 1983). A protein of m.w. 72 500 specific to Mx-bearing cells has been described (Horisberger, Staeheli, and Haller 1983) and an mRNA specific to interferon-treated Mx-carrying cells has been isolated which codes for a similar protein (Staeheli, Colonno, and Cheng 1983). This protein is not induced by interferon-γ. The role of this protein in mediating the antiviral effect of interferon α and β in Mx-bearing cells is not known but it appears that there is a marked inhibition of translation of influenza mRNA in these cells with other steps in the replication cycle remaining unaltered (Meyer and Horisberger 1984). Thus far, this remains the most well characterized virus-specific effect of interferon.

3.9. CONCLUSION

The studies cited here on the antiviral activities of interferons *in vitro* illustrate that the mechanisms of inhibition may be significantly different for different viruses. Moreover, the extent of interferon-induced viral inhibition, although host-cell specific, can vary with the particular molecular species of interferon. An understanding of the relationship between *in vitro* and *in vivo* effects of interferons remains a problem as indirect mechanisms of action are likely to be important *in vivo*. However, *in vitro* observations are relevant in so far as they may indicate relative sensitivities to interferon therapy. Furthermore, an understanding of the molecular mechanisms of action of an interferon type against a specific virus in a particular host cell *in vitro* will be useful in designing and targeting synthetic interferons.

Acknowledgements

B. R. G. Williams is a MRC (Canada) scholar.

3.10. REFERENCES

Aboud M., Shoor, R., and Salzberg, S. (1980). An effect of interferon on the uncoating of murine leukemia virus not related to the antiviral state. *J. Gen. Virol.* **51,** 425–9.

Adams, A. B., Lidin, B., Strander, H., and Cantell, K. (1975). Spontaneous interferon production and Epstein–Barr virus antigen expression in human lymphoid cell lines. *J. Gen. Virol.* **28,** 219–23.

Aguet, M., Belardelli, R., Blanchard, B., Marcucci, F., and Gresser, I. (1982). High affinity binding of ^{125}I-labelled mouse interferon to a specific cell surface receptor. IV. Mouse γ interferon and cholera toxin do not compete for the common receptor site α/β interferon. *Virology* **117,** 541–4.

Ahl, R. and Rump, A. (1976). Assay of bovine interferons in cultures of the porcine cell line IB-RS-2. *Infect. Immunol.* **14,** 603–6.

Allen, G. and Fantes, K. H. (1980). A family of structural genes for human lymphoblastoid (leukocyte-type) interferon. *Nature, Lond.* **287,** 408–11.

Anderson, P., Yip, Y. K., and Vilcek, J. (1982). Specific binding of ^{125}I-human interferon-γ to high affinity receptors on human fibroblasts. *J. Biol. Chem.* **257,** 11301–4.

Arnheiter, H. and Haller, O. (1983). Mx gene control of interferon action: different kinetics of the antiviral state against influenza virus and vesicular stomatitis virus. *J. Virol.* **47,** 626–30.

Aujean, D., Sanceau, J., Falcoff, E., and Falcoff, R. (1979). Location of enhanced ribonuclease activity and of phosphoprotein kinase in interferon-treated mengovirus infected cells. *Virology* **92,** 583–6.

Avery, R. J. Norton, J. D., Jones, J. S., Burke, D. C., and Morris, A. G. (1980). Interferon inhibits transformation of murine sarcoma virus before integration of provirus. *Nature* **288,** 93–5.

Branca, A. A. and Baglioni, C. (1981). Evidence that types I and II interferons have different receptors. *Nature, Lond.* **294,** 768–70.

Brennan, M. B. and Stark, G. R. (1983). Interferon pretreatment inhibits Simian Virus 40 infections by blocking the onset of early transcription. *Cell* **33,** 811–6.

Cayley, P. J., Knight, M., and Kerr, I. M. (1982). Virus-mediated inhibition of the ppp(A2'p)A system and its prevention by interferon. *Biochem. Biophys. Res. Commun.* **104,** 376–82.

Chatterjee, S., Cheung, H. S., and Hunter, E. (1982). Interferon inhibits Sendai virus-induced cell fusion: an effect on cell membrane fluidity. *Proc. Nat. Acad. Sci. USA* **79,** 835–9.

Czarniecki, C. W., Fennie, C. W., Powers, D. B., and Estell, D. A. (1984). Synergistic antiviral and antiproliferative activities of *Escherichia coli* derived human alpha, beta and gamma interferons. *J. Virol.* **49,** 490–6.

Dalton, B. J. and Paucker, K. (1979). Antigenic properties of human lymphoblastoid interferons. *Infect. Immunol.* **23,** 244–8.

Doetsch, P. W., Suhadolnik, R. J., Sawada, Y., Mosca, J. D., Flick, M. B., Reichenbach, N. L., Dang, A. Q., Wu, J. M., Charubala, R., Pfleiderer, W., and Henderson, E. E. (1981). Core (2'–5') oligoadenylate and the cordycepin analog: inhibitors of Epstein–Barr virus induced transformation of human lymphocytes in the absence of interferon. *Proc. Nat. Acad. Sci. USA* **78,** 6699–703.

Dubois, M. F., Vignal, M., Le Cunff, M., and Chany, C. (1983). Interferon inhibits transformation of mouse cells by exogenous cellular or viral genes. *Nature, Lond.* **303,** 433–55.

Eppstein, D. A. and Marsh, Y. V. (1984). Potent synergistic inhibition of herpes simplex virus-2 by 9-[(1,3-dihydroxy-2-propoxy)methyl]guanine in combination with recombinant interferons. *Biochem. Biophys. Res. Commun.* **120,** 66–73.

——, Barnett, J. W., and Marsh, Y. V., Gosselin, G., and Imbach, J.-L. (1983). Xyloadenosine analogue of $(A2'p)_2$ inhibits replication of herpes simplex viruses 1 and 2. *Nature, Lond.* **302,** 723–4.

——, Marsh, Y. V., and Schryver, B. B. (1984). Mechanism of antiviral activity of (XyloA2'p)2XyloA. *Virology* **131,** 341–54.

Esteban, M. (1984). Defective vaccinia virus particles in interferon-treated infected cells. *Virology* **133,** 220–7.

Evinger, M., Rubinstein, M., and Pestka, S. (1981). Antiproliferative and antiviral activities of human leukocyte interferons. *Arch. Biochem. Biophys.* **210,** 319–29.

Fish, E. N., Banerjee, K., and Stebbing, N. (1983). Human leukocyte interferon subtypes have different antiproliferative and antiviral activities on human cells. *Biochem Biophys. Res. Commun.* **112,** 537–46.

Fleischman, W. R. Jr., Georgiades, J. A., Osborne, L. C., Johnson, H. M. (1979). Potentiation of interferon activity by mixed preparations of fibroblast and immune interferon. *Infect. Immunol.* **26,** 248–53.

Gallagher, J. G. and Khoobyarian, N. (1969). Adenovirus susceptibility to interferon: sensitivity of types 2, 7 and 12 to human interferon. *Proc. Soc. Exp. Biol. Med.* **130,** 137–42.

Glasgow, L. A., Hanshaw, J. B., Merigan, T. C., and Petralli, J. K. (1967). Interferon and cytomegalovirus *in vivo* and *in vitro. Proc. Soc. Exp. Biol. Med.* **125,** 843–9.

Goeddel, D. V., Leung, D. W., Dull, T. J., Gross, M., Lawn, R. M., McCandliss, R.,

Seeburg, P. H., Ullrich, A., Yelverton, E., and Gray, P. W. (1981). The structure of eight distinct cloned human leukocyte interferon cDNAs. *Nature, Lond.* **290,** 20–6.

Golgher, R. R., Williams, B. R. G., Gilbert, C. S., Brown, R. E., and Kerr, I. M. (1980). Protein kinase activity and the natural occurrence of 2-5A in interferon-treated EMC virus-infected L-cells. *Ann. N.Y. Acad. Sci.* **350,** 448–56.

Goswami, B. B., and Sharma, O. K. (1984). Degradation of rRNA in interferon-treated vaccinia virus infected cells. *J. Biol. Chem.* **259,** 1371–4.

Gupta, S. L., Holmes, S. L., and Mehra, L. L. (1982). Interferon action against reovirus: activation of interferon-induced protein kinase in mouse L929 cells upon reovirus infection. *Virology* **120,** 495–9.

Haller, O. (1981). Inborn resistance of mice to orthomyxoviruses. *Curr. Top. Microbiol. Immunol.* **92,** 25–52.

Hannigan, G. E., Fish, E. N., and Williams, B. R. G. (1984). Modulation of human interferon-α receptor expression by human interferon-γ. *J. Biol. Chem.* **259,** 8084–6.

—— Gewert, D. R., Fish, E. N., Read, S. E. and Williams, B. R. G. (1983). Differential binding of human interferon-α subtypes to receptors on lymphoblastoid cells. *Biochem. Biophys. Res. Commun.* **110,** 537–44.

Havell, E. A., Yip, Y. K., and Vilcek, J. (1977). Characteristics of human lymphoblastoid (Namalva) interferon. *J. Gen. Virol.* **38,** 51–60.

Hersch, C. L., Brown, R. E., Roberts, W. K., Swryd, E. A., Kerr, I. M., and Stark, G. R. (1984). Simian Virus 40-infected, interferon-treated cells contain 2′, 5′ oligoadenylates which do not activate cleavage of RNA. *J. Biol. Chem.* **259,** 1731–7.

Horisberger, M. A., Staeheli, P., and Haller, O. (1983). Interferon induces a unique protein in mouse cells bearing a gene for resistance to influenza virus. *Proc. Nat. Acad. Sci. USA* **80,** 1910–14.

Hovanessian, A. G., La Bonnardiere, C., Falcoff, E. (1980). Action of murine γ (immune) on β (fibroblast)-interferon resistant L 1210 and embroyonal carcinoma cells. *J. Interferon Res.* **1,** 125–35.

—— Meurs, E., and Montagnier, L. (1981). Lack of systematic correlation between the interferon mediated antiviral state and the levels of 2-5A synthetase and protein kinase in three different types of murine cells. *J. Interferon Res.* **1,** 179–90.

Huleihel, M. and Aboud, M. (1983). Inhibition of retrovirus DNA supercoiling in interferon-treated cells. *J. Virol.* **48,** 120–6.

Jacobsen, H., Krause, D., Friedman, R. M., and Silverman, R. H. (1983). Induction of ppp(A2′p)n A-dependent RNAse in murine JLS-V9R cells during growth inhibition. *Proc. Nat. Acad. Sci. USA.* **80,** 4954–8.

Jay, F. T., Dawood, M. R., Friedman, R. M. (1983). Interferon induces the production of membrane protein deficient and infectivity deficient vesicular stomatitis virions through interference in the virion assembly process. *J. Gen. Virol.* **64,** 707–12.

Johnstone, M. I. and Torrence, P. F. (1984). The role of interferon-induced proteins, double-stranded RNA and 2′5′-oligoadenylate in the interferon mediated inhibition of viral translation. In *Interferons and interferon inducers,* (*ed.* R. M. Friedman). Academic Press, London. (in press)

Lerner, A. M. and Bailey, E. J. (1974). Synergy of 9-β-D-arabinofuranosyladenine and human interferon against herpes simplex virus type I. *J. Infect. Dis.* **130,** 549–52.

Levin, M. J. and Leary, P. L. (1981). Inhibition of human herpesviruses by combination of acyclovir and human leukocyte interferon. *Infect. Immunol.* **32,** 995–9.

Levy, W. P., Shirley, J., Rubinstein, M., Valle, U. D., and Pestka, S. (1980).

Aminoterminal amino acid sequence of human leukocyte interferon. *Proc. Nat. Acad. Sci. USA* **77,** 5102–4.

Lidin, B. and Lamon, E. W. (1982). Antiviral effects of interferon on a somatic cell hybrid between two Burkitt's lymphoma cell lines of different interferon sensitivities. *Infect. Immunol.* **36,** 847–9.

Lin, S. L., Garber, E. A., Wang, E., Caliguiri, L. A., Schellekens, H., Goldberg, A. R., and Tamm, I. (1983). Reduced synthesis of pp 60^{src} and expression of the transformation-related phenotype in interferon-treated Rous sarcoma virus-transformed cells. *Mol. Cell. Biol.* **3,** 1656–64.

Maheshwari, R. K., Banerjee, D. K., Waechter, C. J., Olden, K., and Friedman, R. M. (1980). Interferon treatment inhibits glycosylation of a viral protein. *Nature, Lond.* **287,** 454–6.

Mar, E.-C., Patel, P. C., and Huang, E.-S. (1982). Effect of 9-(2-hydroxyethoxymethyl) guanine on viral-specific polypeptide synthesis in human cytomegalovirus-infected cells. *Am. J. Med.* **73,** 82–5.

Masucci, M. G., Szigeti, R., Klein, E., Gruest, J., Taira, H., Hall, A., Nagata, S., and Weissmann, C. (1980). Effect of interferon-alpha 1 from *E. coli* and some cell functions. *Science* **209,** 1431–4.

Merigan, T. C., Reed, S. E., Hall, T. S., and Tyrrell, D. A. J. (1973). Inhibition of respiratory virus infection by locally applied interferon. *Lancet* **7803,** 561–7.

Merlin, G., Chebath, J., Benech, P., Metz, R., and Revel, M. (1983). Molecular cloning and sequence of partial cDNA for interferon-induced (2′-5′) oligo (A) synthetase mRNA from human cells. *Proc. Nat. Acad. Sci. USA.* **80,** 4904–8.

Meyer, T. and Horisberger, M. A. (1984). Combined action of mouse α and β interferons in influenza virus-infected macrophages carrying the resistance gene Mx. *J. Virol.* **49,** 709–16.

Munoz, A. and Carrasco, L. (1981). Protein synthesis and membrane integrity in interferon-treated HeLa cells infected with encephalomyocarditis virus. *J. Gen. Virol* **56,** 153–62.

—— and —— (1984). Comparison of the antiviral action of different human interferons against DNA and RNA viruses. *FEMS Microb. Lett.* **21,** 105–11.

Nilsen, T. W. and Baglioni, C. (1979). Mechanism for discrimination between viral and host mRNA in interferon-treated cells. *Proc. Nat. Acad. Sci. USA* **76,** 2600–4.

—— Maroney, P. A., and Baglioni, C. (1982*a*). Inhibition of protein synthesis in reovirus-infected Hela cells with elevated levels of interferon-induced protein kinase activity. *J. Biol. Chem.* **257,** 14593–6.

—— —— and —— (1982*b*). Synthesis of (2′-5′) oligoadenylate and activation of an endoribonuclease in interferon-treated HeLa cells infected with reovirus. *J. Virol.* **42,** 1039–45.

Panet, A. and Falk, H. (1983). Inhibition by interferon of herpes simplex virus thymidine kinase and DNA polymerase in infected and biochemically transformed cells. *J. Gen. Virol* **64,** 1999–2006.

Rice, A. P. and Kerr, I. M. (1984). Interferon-mediated, double-stranded RNA dependent protein kinase is inhibited in extracts from vaccinia virus infected cells. *J. Virol.* **50,** 229–36.

——, Roberts, W. K., and Kerr, I. M. (1984). 2-5A accumulates to high levels in interferon-treated, vaccinia virus-infected cells in the absence of any inhibition of virus replication. *J. Virol.* **50,** 220–8.

Riggin, C. H. and Pitha, P. M. (1982). Effect of interferon on the exogenous Friend leukemia virus infection. *Virology* **118,** 202–13.

Rodriguez, J. E., Loepfe, T. R., and Stinski, M. F. (1983). Human cytomegalovirus persists in cells treated with interferon. *Arch. Virol.* **77,** 277–81.

Rubinstein, M., Levy, W. P., Moschera, J., Lai, C. Y., Hershberg, R. D., Bartlett, R.

T., and Pestka, S. (1981). Human leukocyte interferon: isolation and characterization of several molecular forms. *Arch. Biochem. Biophys.* **210,** 307–18.

Samuel, C. E. and Knutson, G. S. (1981). Mechanism of interferon action: cloned human leukocyte interferons induce protein kinase and inhibit vesicular stomatitis virus but not reovirus replication in human annion cells. *Virology* **114,** 302–6.

—— and —— (1982*a*). Mechanism of interferon action. Kinetics of induction of the antiviral state and protein phosphorylation in mouse fibroblasts treated with natural and cloned interferons. *J. Biol. Chem.* **257,** 11791–5.

—— and —— (1982*b*). Mechanism of interferon action. Kinetics of decay of the antiviral state and phosphorylation in mouse fibroblasts treated with natural and cloned interferons. *J. Biol. Chem.* **257,** 11796–801.

—— and —— (1983). Mechanism of interferon action: human leukocyte and immune interferons regulate the expression of different genes and induce different antiviral states in human amnion U cells. *Virology* **130,** 474–84.

Sawai, H., Imai, J., Lesiak, K., Johnstone, M. I., Torrence, P. F. (1983). Cordycepin analogues of 2-5A and its derivatives. Chemical synthesis and biological activity. *J. Biol. Chem.* **258,** 1671–7.

Sen, G. C. and Herz, R. E. (1983). Differential antiviral effects of interferon in three murine cell lines. *J. Virol.* **45,** 1017–27.

Silverman, R. H., Cayley, P. J., Knight, M., Gilbert, C. S., and Kerr, I. M. (1982). Control of the ppp(A2'p)*n*A system in HeLa cells. Effects of interferon and virus infection. *Eur. J. Biochem.* **124,** 131–8.

——, Krause, D., Jacobsen, H., Leisy, S. A., Barlow, D. P., and Friedman, R. M. (1983). 2-5A dependent RNAase levels vary with interferon-treatment, growth rate and cell differentiation. In: *The biology of the interferon system 1983* (ed. E. De Maeyer and H. Schellekens) pp. 189–200. Elsevier, Amsterdam.

Smith, C. A., Wigdahl, B., and Rapp, F. (1983). Synergistic antiviral activity of acyclovir and interferon on human cytomegalovirus. *Antimicrob. Agents Chemother.* **24,** 325–32.

Spector, S. S., Tyndall, M., and Kelley, E. (1982). Effects of acyclovir combined with other antiviral agents on human cytomegalovirus. *Am. J. Med.* **73,** 36–9.

Staeheli, P., Colonno, R. J., and Cheng, Y.-S. E. (1983). Different mRNAs induced by interferon in cells from inbred mouse strains A/J and A2G. *J. Virol.* **47,** 563–7.

Stanwick, T. L., Schinazi, R. F., Campbell, D. E., and Nahmias, A. J. (1981). Combined antiviral effect of interferon and acylovir on herpes simplex virus types 1 and 2. *Antimicrob. Agents Chemother.* **19,** 672–4.

Stewart, W. E. (1979). Mechanisms of antiviral actions of interferons. In *The interferon system* (ed. W. E. Stewart), pp. 202–6. Springer, New York.

Streuli, M., Hall, A., Boll, W., Stewart II, W. E., Nagata, S., and Weissman, C. (1981). Target cell specificity of two species of human interferon-α produced in *Escherichia coli* and of hybrid molecules derived from them. *Proc. Nat. Acad. Sci. USA* **78,** 2848–52.

Tamm, I., Wang, E., Landsberger, F. R., and Pfeffer, L. M. (1982). Interferon modulates cell structure and function. *ICN–UCLA Symp.* **30,** 151–79.

Vacquero, C., Sanceau, J., and Falcoff, R. (1982). Protein synthesis in extracts from interferon-treated mengovirus infected cells. *Biochem. Biophys. Res. Commun.* **107,** 974–80.

Wallach, D. (1983). Quantification of the antiviral effect of interferon by immunoassay of vesicular stomatitis virus proteins. *J. Gen. Virol.* **64,** 2221–7.

——, Fellous, M., and Revel, M. (1982). Preferential effect of γ interferon on the synthesis of HLA antigens and their mRNAs in human cells. *Nature, Lond,* **299,** 833-6.

Weck, P. K., Apperson, S., May, L., and Stebbing, N. (1981*a*). Comparison of the antiviral activity of various cloned human interferon-α subtypes in mammalian cell cultures. *J. Gen. Virol.* **57,** 233–7.

—— ——, Stebbing, N., Gray, P. W., Leung, D., Shepard, H. M., and Goeddel, D. V. (1981*b*). Antiviral activities of hybrids of two major leukocyte interferons. *Nuc. Acids Res.* **9,** 6153–6.

Weissmann, C. (1981). The cloning of interferon and other mistakes. In: *Interferon* (ed. I. Gresser Vol. 3, pp. 101–34. Academic Press, London.

Whitaker-Dowling, P. A., Wilcox, D. K., Widnell, C. C., and Younger, J. S. (1983). Interferon-mediated inhibition of virus penetration. *Proc. Nat. Acad. Sci. USA.* **80,** 1083–5.

Williams, B. R. G. (1983). Biochemical actions. In *Interferon and Cancer* (ed. K. Sikora) pp. 33–52. Plenum Press, New York.

—— Gilbert, C. S., and Kerr, I. M. (1979*a*). The respective roles of the protein kinase and $pppA_2p5'A2'p5'A$-activated endonuclease in the inhibition of protein synthesis by double-stranded RNA in rabbit reticulocyte lysates. *Nuc. Acids Res.* **6,** 1335–50.

—— Golgher, R. R., Brown, R. E., Gilbert, C. S., and Kerr, I. M. (1979*b*). Natural occurrence of 2-5A in interferon-treated EMC virus-infected L-cells. *Nature* **282,** 582–6.

Yonehara, S., Ishii, A., and Yonehara-Takahashi, M. (1983). Cell surface receptor-mediated internalization of interferon: its relation to the antiviral activity of interferon. *J. Gen. Virol.* **64,** 2409–18.

Zerial, A., Hovanessian, A. G., Stefanos, S., Huygen, K., Weiner, G. H., and Falcoff, E. (1982). Synergistic activities of type 1 (α, β) and type II (γ) murine interferons. *Antiviral Res.* **2,** 227–39.

Zoon, K. C., Smith, M. E., Bridgen, P. J., Anfinsen, C. B., Hunkapiller, M. W., and Hood, L. E. (1980). Amino terminal sequence of the major component of human lymphoblastoid interferon. *Science* **207,** 527–8.

4 The regulatory role of interferons in the human immune response

F. R. Balkwill

4.1. INTRODUCTION

The human immune system, as we understand it today, consists of a network of cells that interact in a number of defined circuits to produce a variety of specific effector cells when an antigen perturbs the homeostasis. Regulation of such immune responses between subsets of these cells occurs by direct cellular interaction and the production of a number of biologically active mediators which are collectively called lymphokines.

Interferons (IFNs) are produced during the activation of an immune response and are part of a 'cascade' of lymphokine chemical messages. However, the role of the IFNs in the immune system, and their relationship to the other lymphokines are not yet clearly defined. The aim of this chapter is to present, within the framework of our current knowledge of human immunology, evidence for a role for IFNs in the regulation of the immune response. Such evidence can be currently obtained from three sources: first, a study of the stimuli that induce IFNs *in vitro* in cells of the immune system; second, a study of the effects of purified preparations of the IFNs on immune cell function and surface antigen expression, and third, a study of the role other lymphokines play in controlling the production of, and response to, IFNs.

In order to discuss IFNs in this context it is first necessary to briefly define the effector cells and molecules of the human immune system that will be discussed in this chapter.

4.2. EFFECTOR CELLS OF THE HUMAN IMMUNE SYSTEM

4.2.1. Lymphoid effector cells

There are two main categories of lymphoid cells which can recognize and react against a wide range of antigens and possess immunological memory, T and B cells. B cells are responsible for specific antibody production, T cells can function as direct effector cells against tumours or microbes and also regulate antibody production by B cells.

With the advent of monoclonal antibodies a series of specific antibody markers for T cells have been established, with one antibody reacting with most T cells, OKT3, but not other cell types, and a range of antibodies that are associated with functional subpopulations (for review see Thomas, Rogozinski, and Chess 1983). T cells defined by the OKT4 antibody are mainly helper/inducer cells which induce the differentiation of B cells into antibody forming cells, and precytotoxic T cells into cytotoxic effector cells. T cells defined by the OKT8 antibody are mainly cytotoxic cells and radiosensitive cells important in the suppression of B cell differentiation. These and other T cell surface molecules defined by monoclonal antibodies are thought to be intimately involved in the functions of the cells they define, as the antibodies can block function and proliferation or be strongly mitogenic for resting cells (for review see Thomas *et al.* 1983).

For optimal recognition of specific antigens, T cells also need to interact with the products of the major histocompatibility complex, MHC (for review see Swain 1983). T cells have virtually no spontaneous cytotoxic or other forms of activity but must be activated, usually by exposure to antigens presented in the correct manner by accessory cells such as macrophages and dendritic cells.

Although T cells are able to interact directly with and cause lysis of target cells bearing antigens to which they have been previously sensitized, they can also directly interact with B cells, probably in a MHC restricted fashion (for review see Singer and Hodes 1983), and indirectly influence B cells and other immune cells by their production of lymphokines.

4.2.2. Natural killer cells

Natural killer (NK) cells are non-adherent, non-phagocytic cells that are not dependent on the thymus for maturation. Although they do share a variety of T cell markers, they also share some markers with macrophages and polymorphonuclear leucocytes, PMN (for review see Herberman and Ortaldo 1981). One of the best markers is a morphological one; the majority of the NK activity in peripheral blood mononculear cells, PBMC, is found in a population of large granular leukocytes, LGL, which consist of approximately 5 per cent PBMC. Unlike T cells, NK cells have spontaneous direct cytotoxic activity against a whole range of syngeneic, allogeneic, and xenogeneic cells, do not possess immunological memory, and their cytotoxic activity does not require the expression of MHC.

4.2.3. Macrophages and polymorphonuclear leucocytes

Macrophages and PMN also have natural cytotoxic or cytostatic activity against cells but they have other roles to play in the defence against disease, being phagocytic cells capable of ingesting a variety of particles, such as microbes and IgG antibody-coated tumour cells. Macrophages are also one type of accessory or antigen presenting cell in immune responses (reviewed by Unanue, Beller, Lu, and Allen 1984), the other major cell type with this function being the non-phagocytic dendritic cell (e.g. Van Voorhis, Valinsky, Hoffman, Luban, Hair, and Steinman 1983). An essential step in antigen stimulation is the activation of T helper cells, and this activation is not induced by free antigen but by antigen presented and often processed by an accessory cell that expresses DR (or Class II) MHC glycoproteins on its surface. This expression of DR antigens is subject to positive and negative control, as will be discussed later.

4.3. EFFECTOR MOLECULES OF THE HUMAN IMMUNE SYSTEM

Lymphokines are intercellular connecting molecules, and over 50 regulatory activities have so far been ascribed to them (Hadden and Stewart 1981). It is, at present, not clear whether the many functions ascribed to lymphokines represent a whole series of different molecules or whether a few molecules have multifunctional properties. Apart from IFNs there are two other lymphokines that have been well characterized: IL2 (T cell-derived growth factor) and IL1 (lymphocyte activating factor produced by macrophages). IL2 is a single 15 000 dalton polypeptide (for review see Smith 1984) released from antigen-triggered T cells that is responsible for antigen-stimulated T cell clonal expansion. Resting T cells do not express IL2 receptors until their antigen receptor is triggered; thus the antigen specificity of T cell clonal expansion is ensured. IL1 (lymphocyte activating factor) is a macrophage-derived polypeptide, m.w. 12 000–15 000, that has a variety of amplifying effects on immunological and inflammatory reactions (reviewed by Oppenheim and Gery 1982). Little IL1 is produced by resting monocytes or macrophages but activated lymphocytes can stimulate macrophages directly to produce IL1 by a cell contact-dependent MHC-restricted pathway, or indirectly by the production of other lymphokines. A wide variety of other agents will directly stimulate macrophages to produce IL1.

Biological effects ascribed to IL1 include the augmentation of mitogenesis by certain thymocyte subsets, enhancement of T cell differentiation to express cell surface markers and produce lymphokines, and the promotion of antibody production by B cells. It also appears to act on a number of non-leucocyte cell types and may be identical to endogenous pyrogen, a macrophage product which elicits fever (Rosenwasser, Dinarello, and Rosenthal 1979).

There are many more activities ascribed to lymphokines and a full description of them and the interactions between them is beyond the scope of this chapter. The above examples however give an idea of the potential of such molecules.

Having briefly outlined a basic framework for the human immune system we will now consider the place of IFN in this network, firstly by outlining the stimuli that induce IFN production in cells of the immune system, the characteristics of the producer cells, and the IFN they produce, as summarized in Table 4.1.

Table 4.1. *IFN production by cells of the immune system*

In vitro stimulus	Producer cell type	IFN type produced	References
Mitogen	All subsets of T cells (with or without 'help' from macrophages), monocytes, B cells, null cells	γ, α, small amounts acid-labile α	Epstein 1977 Epstein *et al.* 1980 Wiranowska-Stewart and Stewart 1981 Matsuyama *et al.* 1981 Chang *et al.* 1982 Ratliff *et al.* 1982 O'Malley *et al.* 1982 Cunningham and Merigan 1984
Antigen in immune individuals	OKT_4+ T cells, B cells	γ, acid-labile α	Green *et al.* 1969 Rasmussen and Merigan 1978 Balkwill *et al.* 1983 Cunningham and Merigan 1984
Tumour cells	Non-T, non-B, L.G.L.	α, γ	Trinchieri *et al.* 1977 Trinchieri *et al.* 1978 Timonen *et al.* 1980 Grönberg *et al.* 1983
None (spontaneous)	L.G.L., OKT_4+ T cells	γ, acid-labileα	Fischer and Rubinstein 1983 Cunningham and Merigan 1984 Martinez-Maza *et al.* 1984

4.4. INTERFERON PRODUCTION BY CELLS OF THE IMMUNE SYSTEM

4.4.1. Stimuli of IFN production

Although the first IFNs isolated were stimulated by virus infection of non-immune cells, and IFNs are defined in terms of their antiviral activity (Stewart 1980) many unrelated stimuli are now known to effect IFN production, including some known to initiate immune responses. Stimulation of cells of the immune system to produce IFNs can occur by three other mechanisms distinct from *de novo* virus infection: (i) non-antigen-specific stimulation by agents such as mitogens; (ii) antigen-specific (including of course viral antigen) stimulation; and (iii) tumour cell stimulation. These topics have recently been reviewed in detail by Epstein (1977), Stanton, Langford, and Weigent (1982), and Wilkinson and Morris (1983), and will be described briefly here. Spontaneous production of IFN by peripheral blood cells has also been described and this may be an indication of *in vivo* activation.

Mitogenic stimulation *in vitro* of lymphocytes results in non-specific polyclonal proliferation and antibody and lymphokine production. Mitogens can be specific for lymphocyte subsets, e.g. T cell specific mitogens such as Concanavalin A, Con A, phytohaemagglutinin (PHA), and OKT3 antibodies; and B cell specific mitogens such as lipopolysaccharide (LPS). IFNs are induced by all such mitogens in unfractionated cultured PBMC with a peak of production around 48 h post stimulation. Macrophages have been found to enhance this production (Epstein 1977).

The antigenic stimulus of differences in MHC seen in the primary mixed lymphocyte culture (MLC) also generates IFN production and although a difference in the HLA-DR region between responder and stimulator peripheral blood mononuclear cells (PBMC) gives maximal IFN production, other differences are sufficient for some production (Andreotti and Cresswell 1981).

In vitro responses in donors who possess immunity to a variety of antigens also result in IFN production. Human PBMC from individuals immune to tetanus or diptheria toxiod (Green, Copperbrand, and Kibrick 1969) and viral antigens such as influenza (Balkwill, Griffin, Band, and Beverly 1983), vaccinia (Epstein 1977) and herpes simplex virus, HSV, (Cunningham and Merigan 1984) all produce IFN late (2–6 days) during a proliferative response to the antigen with kinetics of production that are much slower than virus, MLC, or mitogen-induced IFN.

Tumour cell lines can induce HFN production *in vitro* in unsensitized human PBMC as first shown by Trinchieri, Santoni, and Knowles (1977). The kinetics of this production are much faster than that found in the MLC response and differences in the MHC locus between producer and stimulator cells are unnecessary (reviewed by Herberman and Ortaldo 1981). One problem with this experimental system is that several papers have suggested that mycoplasma contamination of tumour cell lines may be responsible for the interferon production (reviewed by Wilkinson and Morris 1983). However, at least some mycoplasma-free tumour cell lines can induce IFN in PBMC (e.g. Grönberg, Kiessling, Masucci, Guevara, Eriksson, and Klein 1983), although to my knowledge there is no report that freshly isolated tumour cells can do this.

Although there is no evidence for *in vivo* production of IFN at the systemic level during normal immune responses, at a cellular level Martinez-Maza, Anderson, Andersson, Britton, and De Ley (1984), using a haemolytic plaque assay to measure IFN production in single cells, found that there are 1–3 IFN-producing cells per 10^3 PBMC in freshly isolated adult or newborn human blood. It is also of interest that ciruclating IFN, generally only found during acute virus infections, can be found in serum from patients with abnormal immune responses states such as systemmic lupus erythematosus (Preble, Black, Friedman, Klippel, and Vilcek 1982) and Kaposi's sarcoma (De Stephano, Friedman, Friedman-Kien, Geodert, Henriksen, Preble, Sonnaband and Vilcek (1982).

4.4.2. Cells that produce IFNs

There is much conflicting evidence concerning which cells of the immune system produce IFN in response to mitogenic or specific antigenic stimulation, the data being confused by the fact that the cell fractionations necessary to resolve the issue are very rarely complete and contamination by small numbers of cells of another type or subtype often occurs. The recent technology of T cell cloning will help resolve this issue but few papers have yet been published.

Whereas all cells of the immune system can produce IFN when infected with virus, the majority of data indicate that lymphocytes and macrophages produce IFNs in response to specific antigen or mitogenic stimulation, (reviewed by Epstein 1977; Stanton *et al.* 1982; Wilkinson and Morris 1983). Among the lymphocyte subclasses T lymphocytes appear to be an important source both when freshly isolated or as T cell clones or hybrids (e.g. Matsuyama, Sugamura, Kawade, and Hinuma 1982). Macrophages generally act as accessory cells in IFN production (Epstein 1977); however monocyte populations have also been shown to produce IFN in response to mitogens (Wiranowska-Stewart and Stewart 1981).

As described in the introduction, T cells comprise a phenotypically and functionally heterogenous population of cells. However IFN production cannot be assigned to one particular subset and the producer cell may depend on the nature of the stimulus. For instance, Chang, Testa, Kung, Perry, Dreskin, and Goldstein (1982) found that OKT4+ T cells, with 'help' from macrophages, produced IFN in response to the mitogenic stimulus of the OKT3 antibody, whereas O'Malley, Nussabaum-Blumenson, Sheedy, and Grossmayer (1982) found that PHA-induced IFN in the absence of monocytes was produced by E+Fc gamma+ OKT11a+ cells that were OKT4− and OKT8− but also possessed a membrane marker for macrophages, OKM1. Epstein and Gupta (1981) fractionated T cells by their receptors for the Fc portion of IgG or IgM, and found that, in the presence of macrophages, T cells with or without either receptor could produce IFN, and Matsuyama *et al.* (1982) found that some but not all human T cell clones with defined cytotoxic ability could generate IFN. Cunningham and Merigan (1984) found that OKT4+ cells from individuals with recurrent herpes labiles were the predominant producers of IFN when stimulated *in vitro* with HSV, but when stimulated with mitogen, all subsets of T lymphocytes were producers. Likewise Kasahara, Hooks, Dougherty, and Oppenheim (1983) found that OKT4+ and OKT8+ cells produced IFN in response to PHA.

B cells may be also involved in antigen-specific production of IFN during the *in vitro* proliferative response to HSV antigens (Rasmussen and Merigan 1978) but the evidence for production during mitogen stimulation is conflicting (Epstein 1977; Wiranowska-Stewart and Stewart 1981).

A different lymphocyte subpopulation is involved in tumour cell-stimulated IFN production. Several studies have confirmed the observations of

Trinchieri *et al.* (1977) that non-T, non-B cells were responsible (reviewed in Herberman and Ortaldo 1981), and using a fluorescein labelled monoclonal interferon antibody, Timonen, Saskeki, Virtanen, and Cantell (1980) morphologically defined the producer cell as a large granular lymphocyte, LGL, and thus probably an NK cell.

A low density subpopulation of PBMC with the morphology of LGL have been found to spontaneously produce high levels of IFN when cultured *in vitro* at high density (Fischer and Rubinstein 1983). OKT4+ cells from individuals with recurrent herpes labiales spontaneously secreted IFN when they were cultured *in vitro* soon after infection (Cunningham and Merigan, 1984), and the same T cell subtype were mainly responsible for the low level of spontaneous IFN production found in freshly isolated blood using a haemolytic plaque assay (Martinez-Maza, *et al.* 1984).

4.4.3. Characterization of IFNs produced by immune cells

To qualify as an interferon, a factor must be a protein which exerts virus non-specific antiviral activity, at least in homologous cells, through cellular processes involving synthesis of both RNA and protein (Stewart 1980).

To date, three main classes of human IFN have been identified; α, β and γ (Stewart 1980), which were originally defined in terms of their stability to pH2, neutralization by polyclonal antisera, and cross-species reactivity. Alpha IFNs, which are acid-stable, comprise many different subtypes of differing cross-species reactivity and biological activity. Beta and γ IFNs are more strictly species-specific and again antigenically distinct, but whereas β IFNs are acid-stable, γ IFNs are acid-labile. Only one gene has so far been identified for either IFN-β or -γ.

Monoclonal antibodies to IFNs have only recently become available and have not been extensively used to characterize IFNs produced during immune reactions. The data described in this section may appear conflicting because of the use of polyclonal antisera to IFNs and often inadequate characterization. It is also important to remember that the crude supernatants contain antigen, other lymphokines, etc., which may affect their biological activity.

It is clear however from a number of sources that the majority of IFN activity produced during mitogen stimulation is acid-labile interferon which is strictly species-specific and not neutralized by polyclonal antisera to α or β IFNs (e.g. Wiranowska-Stewart and Stewart 1981; Matsuyama *et al.* 1982) thus fulfilling the criteria for IFN-γ. However, small amounts of acid-stable IFN-α are also produced after stimulation with various mitogens (Wiranowska-Stewart and Stewart 1981), and an unusual subtype of IFN-α which breaks the 'rules' of classification in being acid-labile but recognized by polyclonal antisera to IFN-α can also be induced under certain conditions, such as stimulation with *C. parvum* (Epstein, Goldblatt, Yu, Dean, Herberman, Oppenheim, Rocklin, and Sugiyama 1980).

In the human mixed lymphocyte culture IFN-γ again appears to be the main IFN produced, although small amounts of pH2-labile cross-species reactive IFN, neutralized by antibodies to IFN-α were also found (Andreotti and Cresswell 1981).

The type of IFN produced during *in vitro* antigen-specific stimulation of cells from immune donors appears to be more variable. In murine systems there is clear data that IFN-γ is produced (for review see Epstein 1977), but in humans there is evidence for the production of α, γ and acid-labile IFN-α, sometimes at the same time. For instance, IFN produced in response to diptheria or tetanus toxoid was stable to pH4 but did not protect mouse, chick or rabbit cells (Green *et al.* 1969); that induced by flu virus antigens in lymphocytes from donors vaccinated against the virus was acid-labile but cross-species reactive and neutralized by IFN-α antibodies (Balkwill *et al.* 1983); and similarly, during *in vitro* proliferative responses of immune donors to HSV, one study showed that IFN produced was an acid-labile IFN-α, little affected by antibody to IFN-γ (Cunningham and Merigan 1984), and another study found IFN actively partially inactivated by pH2 and IFN-α antiserum (Green, Yeh, and Overall 1981).

The situation is of course complicated by the viral nature of some of the stimulating antigens, although production of immune IFN described here is a late (3–7 day) event in the cultures, at the time of peak cell proliferation, unlike that produced during virus infection of cells.

In response to tumour cells, PBMC predominantly produce an acid-stable IFN (e.g. Trinchieri, Santoli, Dee, and Knowles 1978) and the LGL responsible for its production could be stained by a fluorescein-labelled anti-α IFN monoclonal antibody (Timonen *et al.* 1980). Grönberg *et al.* (1983) reported that IFN-γ can also be produced during target/NK cell interactions. Interferon produced spontaneously by T lymphocytes was characterized as IFN-γ (Cunningham and Merigan 1984; Martinez-Maza *et al.* 1984), but IFN produced when PBMC were cultured at high density was a mixture of IFN-γ and acid-labile IFN-α (Fischer and Rubinstein 1983).

Therefore as summarized in Table 4.1, all classes of T and B lymphocytes, LGL, and monocytes are capable of producing IFNs in a manner which is dependent on the inducing stimulus and the constituents of the PBMC population *in vitro*. There is evidence that a small proportion of PBMC are producing IFN *in vivo* in both the newborn and adult with no evidence of a systemmic immune response. It is of interest that a new type of IFN, acid-labile IFN-α, with shared characteristics of both α and γ IFNs, is found quite commonly during stimulus of cells of the immune system, particularly in the late production of IFN to viral antigens by immune individuals (Balkwill *et al.* 1983; Cunningham and Merigan 1984).

The current availability of many more monoclonal antibodies to IFN-α subtypes and IFNs-β and γ should rapidly improve our knowledge of both the type of IFN produced and its cellular source.

4.5. INFLUENCE OF IFN PRODUCTION ON EFFECTOR CELL FUNCTION AND OUTCOME OF THE IMMUNE RESPONSE

The majority of information currently available to help us determine the regulatory role of IFNs in the immune system comes from *in vitro* studies on the effect of exogenous IFNs on cell function. In many cases this work has been done with impure IFN-γ, often contaminated with other lymphokines, or acid-stable IFNs-α and -β that may be inappropriate. Therefore this section will deal mainly with selected recent studies with human cells using pure cell source or recombinant human IFNs.

IFNs can affect cells of the immune system in three major ways: (1) by altering the cell surface, (2) by altering the production and secretion of internal cellular proteins and (3) by enhancing or inhibiting effector cell functional capacity.

4.5.1. Effects of IFNs on the cell surface

The major effect of IFNs on the cell surface is the enhancement of expression of MHC antigens and, as was described in the introduction, the expression of MHC antigens is a vital controlling mechanism of many immune responses. IFN enhancement of MHC expression was first shown by Lindahl, Leary, and Gresser (1973) in murine cells, but has been confirmed on a variety of fresh human cells and cell lines including those of the immune system. For instance, Fellous, Nir, Wallach, Merlin, Rubinstein, and Revel (1982) found that IFN-α enhanced class I MHC antigen expression in lymphoblastoid cell lines and PBMC but had no effect on the expression of class II HLA-DR antigens, and Wallach, Fellous, and Revel (1982) have shown that, in terms of biological activity, IFN-γ is 100-fold more effective than IFN-α in inducing class I antigens. The increase in expression of class I antigens is a result of activation of the gene and of mRNA (Fellous *et al.* 1982). More recent studies (Rosa, Hata, Abadie, Wallach, Fellous, and Revel 1983) have shown that class II HLA-DR mRNA can also be induced by IFNs and once again, preferential induction by IFN-γ occurs, with IFNs-α or -β having only a weak effect.

The effect of IFNs on class II MHC or HLA-DR antigens is of potential importance for the control of immune responses which, as we have already described, are largely MHC restricted. T helper cells recognize exogenous antigens in association with HLA-DR antigens on the surface of antigen presenting cells such as macrophages, dendritic cells and Langerhans cells. It is known that activated T helper cells secrete a lymphokine(s) that induces and enhances HLA-DR expression on macrophages *in vitro* and the evidence would indicate that this lymphokine activity is IFN-γ (reviewed by Unanue *et al.* 1984). Recent studies have confirmed that pure recombinant IFN-γ (but not α or β) will induce expression of HLA-DR antigens on human foetal monocytes (Kelley, Fiers, and Strom 1984) and increase the level of HLA-

DR on human peripheral blood monocytes (Basham and Merigan 1983). Thus activated T cells, by producing IFN-γ, may recruit additional functional antigen-presenting cells to amplify the immune response.

Human vascular endothelial cells may also act as antigen-presenting cells in an HLA restricted manner, and Pober, Gimbrone, Cottran, Reiss, Burakoff, Fiers, and Ault (1983*a*) found that HLA-DR expression, induced in these cells by products of activated T cells, is also induced by recombinant IFN-γ. This inducible expression of HLA-DR in endothelium may be important both for allograft rejection and for recruitment of circulating T cells into the site of an immune response. The same group (Pober, Collins, Gimbrone, Cottran, Gittin, Fiers, Clayberger, Krensky, Burakoff, and Reiss 1983*b*) also found that cloned IFN-γ could induce DR expression on human dermal fibroblasts which could then be functionally recognized by T cells. Again this finding may be important in allograft rejection and antiviral cellular immunity.

Unlike IFN-α or -β, IFN-γ promotes the enhancement of MHC antigens at concentrations that do not inhibit viral replication, and thus IFN-γ production by sensitized T lymphocytes may amplify the antiviral cellular immune response by enhancing HLA-DR expression but allowing viral replication and expression at the cell surface.

The effects of IFN on HLA-DR antigen synthesis and expression may have implications in pathology. Bottazzo, Pajol-Borrell, and Hanfusa (1983) have hypothesized that autoimmunity may involve local aberrant expression of HLA-DR antigens by epithelial cells, thus making them able to present autoantigen to T cells, and that local persistent viral infection and IFN production may be the start of this process.

Another cell surface effect of IFNs on antigen-presenting cells that has recently been described is an enhancement of the number of receptors for the Fc portion of IgG on human monocytes and macrophages (Guyre, Morganelli, and Miller 1983; Perussia, Dayton, Lazarus, Fanning, and Trinchieri 1983). Fc receptors are important for mononuclear cell functions including clearance of immune complexes, phagocytosis, and antibody-dependent cellular cytotoxicity. Guyre *et al.* (1983) showed that natural or recombinant IFN-γ caused a dramatic (10-fold) enhancement of receptor number whereas IFN-α and β had more modest effects, and Perussia *et al.* (1983) found that only purified or recombinant IFN-γ but not IFNs-α or β could induce the Fc receptor. However, *in vivo* intravenous administration of pure lymphoblastoid cell IFN-α (a mixture of α subtypes) caused elevated Fc receptor expression on PBMC (Rhodes, Jones, and Bleehen 1983), and *in vitro* IFN-β increased both Fc receptors and HLA-DR antigens on human monocytes (Rhodes and Stokes 1982).

Another, presumably cell surface, effect on IFNs that may be important in the regulation of immune responses is a reversible alteration in susceptibility to NK cell-mediated lysis, (e.g. Trinchieri and Santoli 1978), and it is possible yet paradoxical that IFNs produced during NK cell–target cell interactions may in fact allow target cells to escape NK cell lysis. Trinchieri, Granato, and Perussia (1981) showed that all three IFN types could protect human cells from NK lysis and that induction of resistance to lysis requires RNA and protein synthesis. As different cell types have different susceptibilities to this effect of IFNs, Trinchieri *et al.* (1981) suggested that this was part of the control mechanism of NK cell activity allowing concentration of activity of cytotoxic cells on targets not protected by IFN.

A specific effect on the human T cell surface was reported by Johnson and Farrar (1983). They found that IFN-γ induced expression of IL2 receptors on T cells and that in the presence of IFN-γ, T cells showed enhanced proliferation to IL2. This may represent one aspect of the amplifying effects of a lymphokine 'cascade', as will be discussed in a further section.

4.5.2. Effects of IFNs on other cellular proteins

When an IFN binds to its cell surface receptor a series of little understood events is set in motion, culminating in the induction, enhanced synthesis, or inhibition of a number of cellular proteins and unusual enzymic activities (for review see Taylor-Papadimitriou 1980). Understanding this process in lymphocytes would give a biochemical basis to the effects described in the preceding and following sections. However few studies have been carried out. Cooper, Fagnani, London, Trepel, and Lester (1982) found that IFN-α or β caused a 30 per cent inhibition of PHA-stimulated protein and DNA synthesis in PBMC, but had no overall effect on protein synthesis in resting lymphocytes. Two-dimensional gel electrophoresis of both resting and stimulated cells showed enhanced synthesis of eight peptides in T lymphocytes 4–6 h after IFN addition, but the role of these proteins and their relationship to MHC antigens and the already characterized IFN-induced enzymes is unknown.

4.5.3. Changes in functional activity of immune cells after IFN exposure

4.5.3.1. Cytotoxicity and antimicrobial activity

Four immune cell types possess the capacity for cytotoxicity: macrophages, PMN, T lymphocytes, and NK cells. IFNs can enhance or induce such activity in all four.

Macrophages and PMN. In the later stages of an infection, or immediately after recovery, T lymphocytes encountering antigens of the infecting organism confer upon macrophages an enhanced ability to kill microbes in a non-specific manner. This 'message' is conducted by a lymphokine which increases the capacity of the macrophages to secrete chemically reactive, incompletely reduced metabolities of molecular oxygen such as hydrogen peroxide (Nakagawara, De Santis, Nogueira, and Nathan 1982). The secretion of hydrogen peroxide and antimicrobial activity are closely correlated and dependent on this lymphokine, known as macrophage activating factor

(MAF) which is also involved in the activation of macrophages to tumour cell cytotoxicity. Several reports have shown functional identity between MAF and IFN-γ (e.g. Shultz and Kleinschmidt 1983, Le, Prensky, Yip, Chang, Hoffman, Stevenson, Balazs, Sudlik, and Vilcek 1983), and Nathan, Murray, Weibe, and Rubin (1983) have shown that pure recombinant IFN-γ enhances both the production of hydrogen peroxide by human macrophages and their ability to kill an intracellular microbial pathogen. IFN-γ was so potent that macrophage peroxide releasing capacity was stimulated to 50 per cent of the maximal value with a geometric mean concentration of 6 pM. This pure IFN-γ has also been shown to activate monocytes to kill tumour cells (Le *et al.* 1983) at concentrations as low as 1 i.u./ml^{-1}, whereas 200-fold more IFN-α or β had no effect. However HuIFN-α enhanced the ability of human macrophages to kill virus-infected, but not uninfected fibroblasts (Stanwick, Campbell, and Nahmias 1980).

As we have described already, IFNs also increase macrophage phagocytic and cytotoxic capacity by enhancing cell surface Fc receptors. Thus IFN-γ has properties of MAF and α and β IFNs also possess similar, but weaker activating potential. We cannot however rule out the possibility that other factors, as yet unidentified, may have MAF activity.

There is little information concerning the effect of purified IFNs on PMN activity. However, Hokland and Berg (1981) showed that low amounts of purified IFN-α potentiated antibody-dependent cellular cytotoxicity of PMN against a range of antibody-coated target cells, particularly when the antibody was present in suboptimal amounts.

Cytotoxic T lymphocytes (CTLs). Exogenous IFNs have two opposing effects on cytotoxic lymphocytes *in vitro:* they inhibit proliferation of the cytotoxic clone, yet enhance the functional capacity of the mature cell (reviewed by De Maeyer and De Maeyer-Guignard 1981; Moore 1983; Zarling 1984). All classes of IFN have been reported to inhibit T cell proliferation to mitogens, antigens, and allogeneic cells, and yet highly purified IFN-α and β strongly augmented CTL responses during a mixed lymphocyte reaction (Heron, Berg, and Cantell 1976).

Natural killer (NK) cells. Human NK cell activity is stimulated by IFNs and IFN inducers both *in vivo* and *in vitro* and IFNs appear to be one of the main regulators of NK cell function (reviewed by Herberman and Ortaldo 1981; Moore 1983; Zarling 1984). Not only do NK cells directly and non-specifically lyse a variety of cells *in vitro,* but they are also capable of secreting IFNs during interaction with their target cells. Highly purified or recombinant α, β, and γ IFNs can stimulate NK cell cytotoxicity against both tumour cell lines and freshly isolated tumour cells. Exogenous IFN appears to increase NK cell activity, not only by recruiting pre-NK cells to lysis, but also by increasing the spectrum of cells lysed, including NK-resistant freshly isolated tumour cells. IFN administration *in vivo* also results in increased NK cell activity. For example, Einhorn, Blomgren, and Strander (1980) showed that low doses of partially pure leucocyte IFN containing a mixture of α subtypes increased NK activity in 43 patients, with peak levels 24 h after administration; prolonged IFN therapy resulted in elevated levels for several months. Not all studies have shown this enhancement in all patients. Lotzova, Savary, Quesoda, Gutterman, and Hersh (1983) found that the effect of a single subtype of recombinant IFN-α on patient NK cell activity was inversely related to pretreatment NK cell activity. Patients with low pretreatment activity showed stimulation whereas those with medium or high basal levels were not augmented and depression of activity was sometimes seen.

4.5.3.2. Antibody production

The majority of data concerning the effects of exogenous IFNs on the B cell function of antibody production come from murine systems and have primarily used α/β or impure lymphokine containing IFN-γ. The conclusions from such data (reviewed by Johnson 1981; Moore 1983; Zarling 1984) are that IFNs can both suppress or enhance primary or secondary antibody responses both *in vivo* and *in vitro,* depending on the time of addition and dose of IFN. More recently Nakamura, Murray, Weibe, and Rubin (1984) have shown that recombinant mouse IFN-γ given to mice at the time of antigen challenge resulted in a two- to five-fold enhancement of antibody formation. These regulatory effects seem to be on the B cells themselves, the lesser differentiated precursors with the greatest capacity for differentiation being the most susceptible to inhibition, and on macrophages, possibly by enhancing HLA-DR expression and hence antigen presentation.

In human model systems *in vitro,* exogenous IFN-α derived from leucocytes or lymphoblastoid cells will both enhance and suppress polyclonal (mitogen stimulated) and antigen specific, antibody production. Low doses of IFN α, added before or during pokeweed mitogen PWM-stimulated Ig synthesis, (Choi, Lim, and Sanders 1981; Harfast, Huddlestone, Casali, Merigan, and Oldstone 1981; Rodriguez, Prinz, Sibbit, Bankhurst, and Williams 1983) but different mechanisms have been proposed to account for this enhancement, e.g. a direct effect on B cells (Harfast *et al.* 1981); an increased B cell response to helper factors produced by radioresistant T cells (Rodriguez *et al.* 1983); or activation of monocytes (Choi *et al.* 1981).

Higher doses of IFNs were inhibitory. Choi *et al.* (1981) suggested that this was due to inhibition of both helper T cell and B cell proliferation. Electrophoretically pure IFN-α was also capable of inhibiting polyclonal Ig synthesis in a T cell independent system. (Fleischer, Atallah, Tasato, Blaese and Greene 1982).

Antibody production by B cells is regulated by helper and suppressor T cells. Evidence from experiments in mice indicate a close relationship between IFN production by T cells and non-specific suppressor T cell activity and regulation of immune response (reviewed by Johnson 1981; Moore

1983). Mitogen-activated human lymphocytes can suppress *in vitro* cell-mediated antibody responses and data from Kadish, Tonsy, Ya, Doyle, and Bloom (1980) suggest that IFNs produced by human T cells may be the principle agents mediating this mitogen suppression. The IFN induced by mitogens in this system and the suppressor effects seen were neutralized by antibodies to IFN-α, but the IFN activity was also pH2-labile. In another study, recombinant IFN-α activated suppressor activity during PWM stimulated polyclonal antibody synthesis (Shnaper, Aune, and Pierce 1983).

4.6. RELATIONSHIP OF IFNS TO OTHER LYMPHOKINES

The cells of the immune system produce a number of biologically active mediators apart from IFNs. It is, at present, not clear whether the many properties ascribed to these lymphokines represent different functions, or whether a few molecules have multifunctional activities. The best characterized lymphokines are the IFNs (α, β, γ, MAF), IL1 (lymphocyte activating factor produced by macrophages), and IL2 (T cell-derived growth factor). Although studies of their interrelationship are at an early stage it is of interest to look at the links between these lymphokines and their place in the 'cascade' of lymphokines evoked during an immune response.

4.6.1. Relationship between IFNs and IL2

From studies so far it appears that IL2 may be involved in the regulation of immune IFN production, and IFN may be required for IL2 induction of active T cells. It is well established that mitogen stimulation of PBMC generates both IL2 and IFN-γ, and Johnson and Farrar (1983) demonstrated that partially purified IFN-γ caused an increase in IL2 receptors. As we have already described, T cell proliferation is dependent on IL2 receptor expression and density, and it would be important if IFN-γ was one of the regulators of this. However, confirmation of such studies with pure IFN-γ is necessary, and we must explain the apparently paradoxical inhibition of T cell proliferation to antigens or mitogens by exogenous IFNs.

Purified IL2 enhanced, or even induced IFN-γ production by mitogen-stimulated lymphocytes (Pearlstein, Palladino, Welte, and Vilcek 1983, Kasahara *et al.* 1983). Both T4+ and T8+ lymphocytes, stimulated with suboptimal doses of mitogens, could be induced by IL2 to produce IFN-γ (Kasahara *et al.* 1983) and these authors suggested that the IFN-γ inducing activity of mitogens and antigens may be due to IL2 production.

Thus the production and roles of IL2 and IFN-γ appear to be intimately linked, and these two lymphokines appear to amplify the regulatory role of each other.

4.6.2. Relationship between IFNs and IL1

Human macrophages, stimulated *in vitro* with endotoxins, produce IL1.

Pretreatment of macrophages with IFN-α or β enhances the production of this lymphokine (Arenzana-Seisdedos and Virelizier 1983) and this result is consistent with the other macrophage-activating effects of the IFNs described previously. This observation is of potential importance both clinically and immunologically. For instance, it has been suggested that IL1 may be identical to endogenous pyrogen, a macrophage product which elicits fever and this gives a possible explanation of the febrile reaction seen when impure and purified IFNs are administered to patients (reviewed by Smedley and Wheeler 1983). IL1 increases IL2 production and promotes T and B cell proliferation and functional capacity (reviewed Oppenheim and Gery 1982) and thus IFN-induced immunostimulation could be related to enhanced IL1 secretion by macrophages.

4.7. CONCLUSIONS

From the description given here of the activities and production of IFNs during *in vitro* immune responses, it is clear that they have similar properties to other well defined lymphokines such as IL1 and IL2, but are more widely produced than these. IL1 is produced by macrophages, IL2 by T cells, but, IFNs can be produced by all classes of PBMC given the appropriate mitogenic or antigen stimulus. A range of IFNs can be induced in immune cells; classical IFN-α or γ, whose properties and effects on immune responses are currently being characterized, and unusual α subtypes, with similarities to IFN-γ, which are largely uncharacterized and have not yet been purified.

IFNs appear to be particularly efficient at amplifying cytotoxicity and antibody production. The basis for such effects is not yet understood but it is likely that changes in the cell surface are of prime importance. For instance, enhancement or even inducement of class II MHC expression would permit more efficient recognition by T cells of antigen-presenting and B cells; enhancement of IL2 receptor expression would permit greater clonal expansion of the T cells; and enhancement of monocyte/macrophage Fc receptor expression permits enhanced phagocytosis and ultimately antigen expression. It is of interest that IFN-γ, primarily a product of T cells, appears to have more powerful cell surface effects than the 'anti-viral' IFNs-α or -β. Also, IFN-γ exerts its effects on DR and Fc receptor expression at concentrations often too low to prevent virus replication, this possibly permitting an enhanced host response to viral antigens.

Effects of IFNs on other immune cell products are less well understood, although important effects have been seen on hydrogen peroxide and IL1 production by phagocytic cells.

Examples are given here in which the effects of the IFNs appear to be self regulatory, in that higher concentrations or different kinetics of exposure are associated with suppression of immune responses. The major mechanism of this immunosuppression appears to be antiproliferative, affecting clonal expansion of T and B cells. IFNs have rarely been shown to inhibit the activities of mature differentiated cells with limited or absent proliferative potential. However, immunosuppression may also be related to cell surface effects. NK cells, in contact with tumour targets, produce IFNs which stimulate and induce NK cell activity but can also protect target cells from lysis. The reason for this paradox is not yet understood but the range of sensitivity seen in different target cells to this protective effect may provide a clue.

Therefore, this chapter has demonstrated how IFNs, free from contamination with other lymphokines, can have important counterbalancing regulatory effects on human immune responses, at least *in vitro,* and how a range of IFNs are naturally produced during such responses. Further understanding of the role of IFNs in the immune response, and their place in the lymphokine cascade, will be achieved when purified recombinant IFNs and a range of monoclonal antibodies are more widely available. Such knowledge is important, not only to the study of immunology, but also to our understanding of the role that IFNs, as biological response modifiers, can play in the therapy of human disease.

Acknowledgement

The author wishes to thank Dr. Peter Beverley for helpful discussion and criticism.

4.8. REFERENCES

Andreotti, P. E. and Cresswell, P. (1981). HLA control of interferon production in the human lymphocyte culture. *Hum. Immunol.* **3,** 109–20.

Arenzana-Seisdedos, F. and Virelizier, J.-L. (1983). Interferons as macrophage-activating factors II. Enhanced secretion of interleukin 1 by lipopolysaccharide-stimulated human monocytes. *Eur. J. Immunol. 13,* 437–40.

Balkwill, F. R., Griffin, D. B., Band, H. A., and Beverly, P. C. L. (1983). Immune human lymphocytes produce an acid-labile α-interferon. *J. Exp. Med.* **157,** 1059–63.

Basham, T. Y. and Merigan, T. C. (1983). Recombinant interferon-gamma increases HLA-DR synthesis and expression. *J. Immunol.* **130,** 1491–94.

Bottazzo, G. F., Pujol-Borrell, R., and Hanafusa, T. (1983). Role of aberrant HLA-DR expression and antigen presentation in induction of endocrine autoimmunity. *Lancet,* **ii,** 115–8.

Chang, T.-W., Testa, D., Kung, P. C., Perry, L., Dreskin, H. J., and Goldstein, G. (1982). Cellular origin and interactions involved in gamma-interferon production induced by OKT3 monoclonal antibody. *J. Immunol.* **128,** 585–9.

Choi, Y. S., Lim, F. K. H., and Sanders, F. K. (1981). Effect of interferon alpha or pokeweed mitogen-induced differentiation of human peripheral blood B lymphocytes. *Cell. Immunol.* **64,** 20–8.

Cooper, H. L., Fagnani, R., London, J., Trepel, J., and Lester, E. R. (1982). Effect of interferons on protein synthesis in human lymphocytes: enhanced synthesis of eight specific peptides in T cells and activation-dependent inhibition of overall protein synthesis. *J. Immunol.* **128,** 828–33.

Cunningham, A. L. and Merigan, T. C. (1984). Leu-3+ T cells produce gamma-interferon in patients with recurrent herpes labialis. *J. Immunol.* **132,** 197–202.
De Stephano, E., Friedman, R. M., Friedman-Kien, A. E., Goedert, J. J., Henriksen, D., Preble, O. T., Sonnabend, J. A., and Vilcek, J. (1982). Acid labile human leucocyte interferon in homosexual men with Kaposi's sarcoma and lymphadenopathy. *J. Infect. Dis.* **146,** 451–5.
Einhorn, S., Blomgren, H., and Strander, H. (1980). Interferon and spontaneous cytotoxicity in man. Enhancement of spontaneous cytotoxicity in patients receiving human leukocyte interferon. *Int. J. Cancer* **26,** 419–28.
Epstein, L. B. (1977). Mitogen and antigen induction of interferon *in vitro* and *in vivo*. *Tex. Rep. Biol. Med.* **35,** 42–56.
—— and Gupta, S. (1981). Human T-lymphocyte subset production of immune (gamma) interferon. *J. Clin. Immunol.* **1,** 186–94.
—— Goldblatt, D., Yu, K., Dean, J. H., Herberman, R. B., Oppenheim, J. J., Rocklin, R. E., and Sugiyama, M. (1980). Mediator induction in human mononuclear cells by strains of corynebacterium parvum. In *Biochemical characterisation of lymphokines,* (ed. A. De Weck, F. Kristensen, and P. Landy) pp. 353–8. Academic Press, New York.
Fellous, M., Nir, U., Wallach, D., Merlin, G., Rubinstein, M., and Revel, M. (1982). Interferon-dependent induction of mRNA for the major histocompatibility antigens in human fibroblasts and lymphoblastoid cells. *Proc. Nat. Acad. Sci. USA* **79,** 3082–6.
Fischer, D. G. and Rubinstein, M. (1983). Spontaneous production of interferon-gamma and acid-labile interferon-alpha by subpopulations of human mononuclear cells. *Cell. Immunol.* **81,** 426–34.
Fleischer, T. A., Attallah, A. M., Tasato, G., Blaese, R. M., and Greene, W. C. (1982). Interferon-mediated inhibition of human polyclonal immunoglobulin synthesis. *J. Immunol.* **129,** 1099–103.
Green, J. A., Copperband, S. R., and Kibrick, S. (1969). Immune specific induction of interferon production in cultures of human blood lymphocytes. *Science* **164,** 1415–17.
—— Yeh, T.-J., and Overall, J. C. Jr. (1981). Sequential production of IFN-alpha and immune-specific IFN-gamma by human mononuclear leukocytes exposed to herpes simplex virus. *J. Immunol.* **127,** 1192–6.
Grönberg, A., Kiessling, R., Masucci, G., Guevara, L. A., Eriksson, E., and Klein, G. (1983). Gamma interferon (IFN-γ) produced during effector and target interactions renders target cells less susceptible to NK-cell-mediated lysis. *Int. J. Cancer* **32,** 609–16.
Guyre, P. M., Morganelli, P. M., and Miller, R. (1983). Recombinant immune interferon increases immunoglobulin G Fc receptors on cultured human mononuclear phagocytes. *J. Clin. Invest.* **72,** 393–7.
Hadden, J. W., and Stewart, W. E., eds. (1981). *The lymphokines: biochemistry and biological activity*. Humana Press, New Jersey.
Harfast, B., Huddlestone, J. R., Casali, P., Merigan, T. C., and Oldstone, M. B. A. (1981). Interferon acts directly on human B lymphocytes to modulate immunoglobulin synthesis. *J. Immunol.* **127,** 2146–50.
Herberman, R. B. and Ortaldo, J. R. (1981). Natural killer cells: their role in defenses against disease. *Science* **214,** 24–30.
Heron, I., Berg, K., and Cantell, K. (1976). Regulatory effect of interferon on T cells *in vitro*. *J. Immunol.* **117,** 1370–3.
Hokland, P. and Berg, K. (1981). Interferon enhances the antibody-dependent cellular cytotoxicity (ADCC) of human polymorphonuclear lymphocytes. *J. Immunol.* **127,** 1585–8.

Johnson, H. M. (1981). Effect of interferon on antibody formation. *Tex. Rep. Biol. Med.* **41,** 411–9.
—— and Farrar, W. L. (1983). The role of a gamma interferon-like lymphokine in the activation of T cells from expression of interleukin 2 receptors. *Cell. Immunol.* **75,** 154–9.
Kadish, A. S., Tansey, F. A., Yu, G. S. M., Doyle, A. T., and Bloom, B. R. (1980). Interferon as a mediator of human lymphocyte suppression. *J. Exp. Med.* **151,** 637–50.
Kasahara, T., Hooks, J. J., Dougherty, S. F., and Oppenheim, J. J. (1983). Interluekin 2-mediated immune interferon (IFN-γ) production by human T cells and T cell subsets. *J. Immunol.* **130,** 1784–9.
Kelley, V. E., Fiers, W., and Strom, T. B. (1984). Cloned human interferon-gamma, but not interferon-beta or -alpha induces expression of HLA-DR determinants by fetal monocytes and myeloid leukemic cell lines. *J. Immunol.* **132,** 240–5.
Le, J., Prensky, W., Yip, Y. K., Chang, Z., Hoffman, T., Stevenson, H. C., Balazs, I., Sudlik, J. R., and Vilcek, J. (1983). Activation of human monocyte cytotoxicity by natural and recombinant immune interferon. *J. Immunol.* **131,** 2821–6.
Lindahl, P., Leary, P., and Gresser, I. (1973). Enhancement of interferon of the expression of surface antigen on murine leukemia L1210 cells. *Proc. Nat. Acad. Sci. USA.* **70,** 2785–8.
Lotzova, E., Savary, C. A., Quesada, J. R., Gutterman, J. V., and Hersh, E. M. (1983). Analysis of natural killer cell cytoxicity of cancer patients treated with recombinant interferon. *J. Nat. Cancer Inst.* **71,** 903-10.
Matsuyama, M., Sugamura, K., Kawade, Y., and Hinuma, Y. (1982). Production of immune interferon by cytotoxic T cell clones. *J. Immunol.* **129,** 450–1.
Martinez-Maza, O., Anderson, U., Andersson, J., Britton, S., and De Ley, M. (1984). Spontaneous production of interferon-gamma in adult and newborn humans. *J. Immunol.* **132,** 251–5.
Moore, M. (1983). Interferon and the immune system 2: effect of IFN on the immune system. In *Interferons: from molecular biology to clinical application. SGM Symposium 35* (ed. D. C. Burke and A. G. Morris) pp. 181–210. Cambridge University Press, Cambridge.
Nakagawara, A., De Santis, N. M., Nogueira, N., and Nathan, C. F. (1982). Lymphokines enhance the capacity of human monocytes to secrete reactive oxygen intermediates. *J. Clin. Invest.* **70,** 1042–68.
Nakamura, M., Manser, T., Pearson, G. D. N., Daley, M. J., and Daley, M. J. (1984). Effect of IFN-gamma on the immune response *in vivo* and on gene expression *in vitro*. *Nature* **307,** 381–2.
Nathan, C. F., Murray, H. W., Wiebe, M. E., and Rubin, B. Y. (1983). Identification of interferon-gamma as the lymphokine that activates human macrophage oxidative metabolism and antimicrobial activity. *J. Exp. Med.* **158,** 670–89.
O'Malley, J. A., Nussabaum-Blumenson, A., Sheedy, D., Grossmayer, B. J., and Ozer, H. (1982). Identification of the T cell subset that produces human gamma interferon. *J. Immunol.* **128,** 2522–6.
Oppenheim. J. J. and Gery, I. (1982). Interleukin 1 is more than an interleukin. *Immunol. Today* **3,** 113–9.
Pearlstein, K. T., Palladino, M. A., Welte, K., and Vilcek, J. (1983). Purified human interleukin-2 enhances induction of immune interferon. *Cell. Immunol.* **80,** 1–9.
Perussia, B., Dayton, E. T., Lazarus, R., Fanning, V., and Trinchieri, G. (1983). Immune interferon induces the receptor for monomeric IgG1 on human monocyte and myeloid cells. *J. Exp. Med.* **158,** 1091–113.
Pober, J. S., Gimbrone, A. Jr., Cotran, R. S., Reiss, C. S., Burakoff, S. J., Fiers, W., and Ault, K. A. (1983*a*). Ia expression by vascular endothelium is inducible by

activated T cells and by human gamma interferon. *J. Exp. Med.* **157,** 1339–53.
—— Colins, T., Gimbrone, M., Cotran, R., Gitlin, J., Fiers, W., Clayberger, C., Krensky, A., Burakoff, S., and Reiss, C. (1983*b*). Lymphocytes recognise human vascular endothelial and dermal fibroblast Ia antigens induced by recombinant immune interferon. *Nature* **305,** 726–9.

Preble, O. T., Black, R. J., Friedman, R. M., Klippel, J. H., and Vilcek, J. (1982). Systemic lupus erythematosus: presence in human sera of an unusual acid-labile leucocyte interferon. *Science* **216,** 429–31.

Rasmussen, L. and Merigan, T. C. (1978). Role of T lymphocytes in cellular immune responses during herpes simplex virus infection in humans. *Proc. Nat. Acad. Sci. USA* **75,** 3957–61.

Rhodes, J. and Stokes, P. (1982). Interferon-induced changes in the monocyte membrane: inhibition by retinol and retinoic acid. *Immunology* **45,** 531–53.
—— Jones, D. H., and Bleehen, N. M. (1983). Increased expression of human monocyte HLA-DR antigens and Fcgamma receptors in response to human interferon *in vivo*. *Clin. Exp. Immunol.* **53,** 739–43.

Rodriguez, M. A., Prinz, W. A., Sibbitt, W. L., Bankhurst, A. D., and Williams, R. C. Jr. (1983). Alpha-interferon increases immunoglobulin production in cultured human mononuclear leukocytes. *J. Immunol.* **130,** 1215–19.

Rosa, F., Hata, D., Abadie, A., Wallach, D., Revel, M., and Fellous, M. (1983). Differential regulation of HLA-DR mRNAs and cell surface antigens by interferon. *EMBO J.* **2,** 1585–9.

Rosenwasser, L. J., Dinarello, C. A., and Rosenthal, A. S. (1979). Adherent cell function in murine T-lymphocyte antigen recognition. *J. Exp. Med.* **150,** 709–13.

Schnaper, H. W., Aune, T. M., and Pierce, C. W. (1983). Suppressor T cell activation by human leukocyte interferon. *J. Immunol.* **131,** 2301.

Shultz, R. M. and Kleinschmidt, W. J. (1983). Functional identity between murine γ interferon and macrophage activating factor. *Nature* **305,** 239–40.

Singer, A. and Hodes, R. J. (1983). Mechanisms of T cell–B cell interaction. *Ann. Rev. Immunol.* **1,** 211–41.

Smedley, H. M. and Wheeler, T. (1983). Toxicity of interferon. In *Interferon and Cancer* (ed. K. Sikora) pp. 203–10. Plenum Press, New York.

Smith, K. A. (1984). Interleukin 2. *Ann. Rev. Immunol.* (in press).

Stanton, G. J., Langford, M. P., and Weigent, D. A. (1982). Cell types involved in production of interferon by leukocytes. *Tex. Rep. Biol. Med.* **41,** 84–120.

Stanwick, T. L., Campbell, D. E., and Nahmias, A. J. (1980). Spontaneous cytotoxicity mediated by human monocyte-macrophages against human fibroblasts infected with herpes simplex virus-augmentation by interferon. *Cell. Immunol.* **53,** 413–6.

Stewart, W. E. (1980). Interferon nomenclature. *Nature* **286,** 110.

Swain, S. (1983). T cell subsets and the recognition of MHC class. *Immunol. Rev.* **74,** 129–42.

Taylor-Papadimitriou, J. (1980). Effects of interferons on cell growth and functions. In *Interferon 2* (ed. I. Gresser) pp. 13–46. Academic Press, London.

Thomas, Y., Rogozinski, L., and Chess, L. (1983). Relationship between human T cell functional heterogeneity and human T cell surface molecules. *Immunol. Rev.* **74,** 113–28.

Timonen, T., Saksela, E., Virtanen, I., and Cantell, K. (1980). Natural killer cells are responsible for the interferon production induced in human lymphocytes by tumor cell contact. *Eur. J. Immunol.* **10,** 422–7.

Trinchieri, G. and Santoli, D. (1978). Antiviral activity induced by culturing lymphocytes with tumour-derived or virus-transformed cells. Enhancement of human natural killer cell activity by interferon and antagonistic inhibition of

susceptibility of target cells to lysis. *J. Exp. Med.* **147,** 1314–33.
—— Granato, D. and Perussia, B. (1981). Interferon-induced resistance of fibroblasts to cytolysis mediated by natural killer cells: specificity and mechanism. *J. Immunol.* **126,** 335–40.
—— Santoli, D., and Knowles, B. B. (1977). Tumour cell clines induce interferon in human lymphocytes. *Nature* **270,** 611–13.
—— Santoli, D., Dee, R. R., and Knowles, B. B. (1978). Anti-viral activity induced by culturing lymphocytes with tumor-derived or virus-transformed cells. *J. Exp. Med.* **1477,** 1299–312.
Unanue, E. R., Beller, D. I., Lu, C. Y., and Allen, P. M. (1984). Antigen presentation: comments on its regulation and mechanism. *J. Immunol.* **132,** 1–5.
Van Voorhis, W. C., Valinsky, J., Hoffman, E., Luban, J., Hair, L. S., and Steinman, R. N. (1983). The relative efficiency of human monocytes and deridritic cells as accessory cells for T cell recognition. *J. Exp. Med.* **158,** 174–91.
Waksman, B. H. (1979). Overview: biology of the lymphokines. In *Biology of the lymphokines* (ed. S. Cohen, E. Ack, and J. J. Oppenheim) pp. 585–616. Academic Press, London.
Wallach, D., Fellous, M., and Revel, M. (1982). Preferential effect of gamma interferon on the synthesis of HLA antigens and their mRNAs in human cells. *Nature* **299,** 240–5.
Wilkinson, M. and Morris, A. G. (1983). Interferon and the immune system 1: induction of interferon by stimulation of the immune system. In *Interferons: from molecular biology to clinical application. SGM Symposium 35* (ed. D. C. Burke and A. G. Morris), pp. 149–79. Cambridge University Press, Cambridge.
Wiranowska-Stewart, M. and Stewart, W. E. II. (1981). Determination of human leukocyte populations involved in production of interferons alpha and gamma. *J. Interferon Res.* **1,** 233–44.
Zarling, Z. M. (1984). Effects of interferon and its inducers on leucocytes and their immunologic functions. In *Handbook of Experimental Pathology 71* (ed. P. E. Came and W. A. Carter), pp. 403–31. Springer, Berlin.

5 Interferons as regulators of cell growth and differentiation

Joyce Taylor-Papadimitriou and Enrique Rozengurt

5.1. HISTORICAL INTRODUCTION

The first report describing an effect of an interferon on cell proliferation appeared in 1962, when the inhibition of growth of mouse embryo fibroblasts by preparations of mouse interferon was described by Paucker and colleagues (Paucker, Cantell, and Henle 1962). However, it was not until purified preparations of interferons were available that it was possible to attribute the growth inhibitory effects observed to the interferon molecules themselves. It is now quite clear, using purified interferons prepared from natural sources and by recombinant DNA technology that interferons of the α, β, and γ classes can inhibit cell proliferation in a wide variety of cell types.

The first indications that interferons can affect other cell functions came from work investigating the effects of interferon on animal tumours. Although these experiments were done initially using tumours of known viral aetiology it soon became obvious that the growth of chemically induced, transplantable, and spontaneous tumours of non-viral aetiology could also be inhibited by interferon (Gresser and Tovey 1978). In view of the observed inhibitory effect of interferons on cell proliferation, a plausible explanation for its antitumour effect was that it directly inhibited the growth of the tumour cells. Moreover, Chany and colleagues had also shown that an MSV-transformed cell line showed a reversion to the normal phenotype after prolonged culture with interferon (Chany and Vignal 1968), so that the *in vivo* inhibition of tumour growth could also be related to this phenomenon. However, it was then found that a strain of L1210 cells whose growth was not inhibited by interferon *in vitro* could be inhibited *in vivo* (Gressner, Maury, and Brouty-Boyé 1972). This observation showed clearly that factors other than a direct inhibition of tumour cell growth or alteration of tumour cell phenotype were involved in the antitumour effect of interferon, and led to the demonstration that interferons can dramatically affect the functions of the effector cells of the immune system (Balkwill, Chapter 4, this volume).

Although the regulation of differentiation and differentiated function by interferons was first observed with the effector cells of the immune system, the effects are not restricted to these cells, but extend to other cell types exhibiting a wide variety of functions (Taylor-Papadimitriou 1980, 1983). Moreover, many of these effects have been observed with interferons purified to homogeneity. The interferons therefore represent a unique family of molecules, naturally produced by cells, which can inhibit cell growth and modulate cell function. In this article we will consider (1) some of the general features of the effects of interferons on cell growth and function; (2) some of the molecular responses of the interferon-treated cell; (3) approaches to studying the mechanism underlying the inhibition of cell proliferation induced by interferons; and (4) the importance of the receptor interaction. Most of the studies discussed have been performed with cultured cells, mainly because the effects of interferons in multicellular organisms, where cell interactions may obscure a primary response, are difficult to interpret.

5.2. INHIBITION OF CELL GROWTH BY INTERFERONS

Activities of interferons have been traditionally defined in terms of antiviral units, and the concentration of interferon required to affect the growth of cultured cells is between 0.2 and 10 000 i.u. ml^{-1} depending on the cell type. With the advent of purified interferons it is possible to give these numbers definition in terms of mg of protein, and using interferon from Namalwa cells (HuIFN-αN) with a specific activity of 3×10^8 i.u. mg^{-1} (Fantes and Allen 1981) the effective dose range is roughly 2 pg to 50 ng ml^{-1}. In short-term culture interferons have been found to be cytostatic rather than cytotoxic, with the exception of the Daudi cell, which may be killed by low concentrations of HuIFN-α.

Although much of the work on interferon-induced growth inhibition has been done with cell lines, many of which were derived from tumours, there is data showing that normal diploid cells (fibroblasts, mammary epithelial cells, lymphocytes) are also inhibited. If a range of cell lines deriving from tumours of the same type are considered then a wide range of sensitivity is noted. This has been seen with osteosarcoma, and with lymphoblastoid, breast carcinoma, and melonoma lines. The lymphoblastoid cell line Daudi derived from a Burkitts lymphoma, and the BT20 cell line derived from a primary breast cancer can be significantly inhibited by less than 1 i.u. or approximately 3 pg (per ml) of interferon. However, cell lines from ostensibly similar tumours may require 1000 times this amount to show an effect on growth. It has been suggested that tumour cells are more sensitive to interferon's growth inhibitory effect than normal cells. While the most sensitive cells have so far been found to be tumour cell lines, resistant lines from similar tumours are less sensitive than the normal cell from which they derive. Thus, while *in vivo* tumour cells may be more sensitive because of the indirect effects of the immune system, it is not possible to generalize with cultured cells.

In asking what makes one cell sensitive to interferon's growth inhibitory effect and another resistant, we are beginning to approach the question of mechanism of action, which we will discuss later. However, one obvious reason for resistance can be mentioned, namely, the lack of membrane receptors (Aguet 1980). It is also important to remember that there are three classes of interferons and there may be tissue specificity with, for example, lymphoblastoid cells being more sensitive to growth inhibition by α interferons and osteosarcoma cells to β interferons. Accurate comparisons of growth inhibitory activity of the different interferons on a mg of protein basis are only just beginning to be made and comparisons on the basis of antiviral units are difficult to interpret. However, since there are three classes of interferons, and several subspecies of IFN-α, it seems possible and even probable that these will show differences in their effectiveness as growth inhibitory agents in different cell types. One observation which is potentially of great interest is that α and β interferons may synergize with γ interferons in inhibiting cell or tumour growth (Fleischman, Kleyn, and Bron 1980). In the mouse and human systems at least, γ interferon does not interact with the receptor used by α and β interferons (Ankel, Drishamurti, Besancon, Stefanos and Falcoff 1980; Branca and Baglioni 1981) and may induce a set of host reactions different from those induced by the α and β interferons.

5.3. INTERFERON EFFECTS ON DIFFERENTIATION AND DIFFERENTIATED FUNCTION

At first sight, the list of cell functions which are affected by interferon appear to be extremely diverse, and some may be enhanced and others inhibited (Taylor-Papadimitriou 1980, 1983). If, however, we distinguish between the *acquirement* of a differentiated function (i.e. the process of differentiation itself) and the expression of a function by a differentiated cell, it becomes clear that it is the development of a differentiated cell from an undifferentiated cell which is inhibited (for examples see Table 5.1), while the expression of a differentiated function by an already differentiated cell is enhanced (see Table 5.2). Although the molecular mechanisms underlying

Table 5.1. *Inhibition of differentiation by interferon*[a]

Production of antibody forming cells
Delayed hypersensitivity reaction
DMSO-induced differentiated in Friend cells
Insulin-induced differentiation of 3T3 cells to adipocytes
Maturation of human monocytes to macrophages

[a]For references describing phenomena described in Tables 5.1–4, see reviews by Taylor-Papadimitriou (1980, 1983).

these phenomena are certainly not clear, it is likely that they may involve common factors. Thus, the increased expression of a differentiated function may be related to the inhibitory effect on cell growth, or to an effect of interferon on a common mechanism operative in the differentiated cell which

Table 5.2. *Differentiated functions enhanced by interferons*

Phagocytosis by macrophages
Cytotoxicity of sensitized lymphocytes
Activity of natural killer cells
Production of antibody
IgE-mediated histamine release by basophils
Synthesis of prostaglandins
Synthesis of ketosteroids in adrenal cells
Beat frequency of myocardial cells
Excitability of cultured neurones
Uptake of iodide by thyroid cells
Expression of Fc receptors
Expression of histocompatibility antigens in mouse and human cells
Expression of Carcinoembryonic antigen

Table 5.3. *Inhibition of induced protein synthesis by interferons*

Mitogen-induced thymidine kinase in L1210 cells
Steroid-inducible glycerol-3-phosphate
Steroid-inducible tyrosine aminotransferase in hepatoma cells
Induction of ornithine decarboxylase activity in fibroblasts by serum growth factors and tumour promoters

could result for example in stabilization of certain messages. In considering the inhibitory effects of interferon on the differentiation process, it is important to note that interferons inhibit not only cell growth and tumourigenicity but also the induction of certain proteins, examples of which are listed in Table 5.3.

Differentiation may involve cell division, the production of specific mitogenic factors (e.g. proteins) and of surface receptors for these factors (glycoproteins), and may require the synthesis of specific messages by the differentiating cell in response to the factors. All of these stages could be affected by interferon, and where inhibition of differentiation is observed, an effect on any one or all of them could be involved.

The most widely studied system, apart from those involving haemopoietic cells (Balkwill, Chapter 4, this volume) is the differentiation to adipocytes which can be induced in 3T3-L1 cells (Keay and Grosberg 1980). In this case specific enzymes related to lipid synthesis are induced during the differentiation process and interferon inhibits their synthesis. Now that other cell systems for studying differentiation are available (e.g. keratinocytes, mammary epithelium, muscle cells) it will be important to test the effect of interferons on their differentiation. It may be that effects of interferon on differentiation and differentiated function seen *in vitro* reflect an important *in vivo* role (as is almost certainly true for the effector cells of the immune system). It may also be true however that the toxic symptoms experienced by patients receiving interferon (e.g. tachycardia, pyrexia) may be related to interferon's effects on differentiation or on differentiated function.

5.4. RESPONSES OF THE INTERFERON-TREATED CELL

Figure 5.1 gives a general outline of the major changes seen in cells treated with an interferon. The ligand first interacts with specific membrane receptors and thus generates signals which result in (1) physical, chemical and

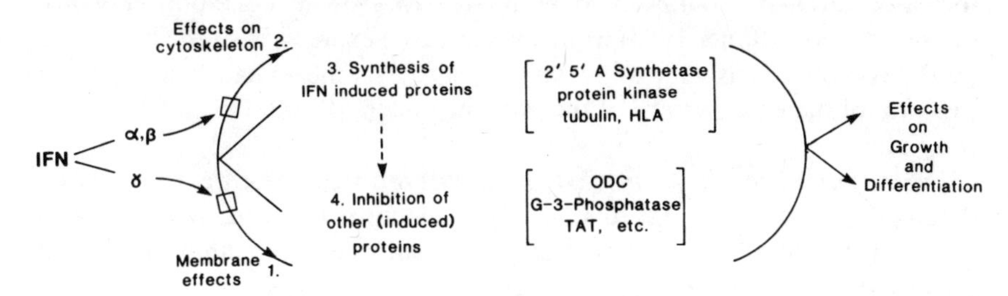

Fig. 5.1. Effects of interferons on cells which may be involved in inhibition of cell growth.

functional changes in the membrane, (2) changes in various components of the cytoskeleton (Bourgeaude, Rousset, Paulin, and Chany 1981), (3) the induction of a range of new proteins, and (4) inhibition of certain inducible proteins. Table 5.4 lists the changes which have been seen in the membranes,

Table 5.4. *Changes in cell membranes induced by interferons*

Increased expression of surface antigens
Increased binding of lectins
Increased net negative charge
Increase in relative proportion of saturated acyl side chains in membrane phospholipids
Increase in intramembrane particles
Inhibition of binding of cholera toxin and TSH
Decreased exposure of oligosaccharide moieties of some gangliosides
Stimulation of adenyl cyclase activity (in certain cell types) and infectivity of virus particles budding from cells
Increase in membrane rigidity
Increased association of actin with the cell membrane
Inhibition of movement of membrane receptors

some of which relate to changes in the cytoskeleton. For example, the decreased movement of interferon treated cells and of membrane receptors is probably related to the increased association of actin with the cell membrane (Wang, Pfeffer, and Tamm 1981). The increased rigidity of the cell membrane (Pfeffer, Landsbergery, and Tamm 1981) could well relate to the growth inhibitory effect, as could the specific inhibition of induced protein synthesis referred to earlier (Table 5.3).

The new proteins which are induced in the interferon-treated cell (Taylor 1964) include 2-5A synthetase and a protein kinase, both of which are dependent on dsRNA for activity and are known to be involved in interferon's antiviral effect (for reviews see McMahon and Kerr 1983; Williams and Fish, this volume, Chapter 2). Whether they are also involved in the effects on cell growth and function is not clear and will be discussed below. Other proteins and their corresponding messages, however, are also induced, and some of these could be involved in the regulatory effects. The increased level of tubulin could be related to growth regulation (Fellous, Ginsberg, and Littauer 1982) while the increased expression of the histocompatibility antigens is almost certainly related to the enhancement of the function of the effector cells of the immune system (See Balkwill, Chapter 4, this volume).

From the above it is clear that the interferon-treated cell exhibits many differences in structure and function from the untreated cell. What remains unclear is which of these changes play a crucial role in the inhibition of cell growth and in the effects on cell function.

5.5. MECHANISMS IN INTERFERON-INDUCED GROWTH INHIBITION

There are several ways to approach the question of how interferons inhibit cell proliferation and here we will consider the data which have been accumulated using three approaches:

1. One approach is to study the changes induced by interferon in cells which are sensitive to its growth inhibitory effect and compare them to those seen in mutant cell lines which are resistant to this action of interferon.

2. A similar stretegy involves examining the sensitivity to interferon of mutant cell lines deficient in certain functions to see if the mutated function is necessary to growth inhibition.

3. A very different approach, which we ourselves have taken, is to examine the effects of interferons on the mitogenic signal in a cell system where some of the molecular events involved in the cell's response to growth factors are known; for these studies it is obviously desirable to use a synchronized system.

5.5.1. Studies with cell lines resistant to interferon's growth inhibitory effect

Cells resistant to interferon may be defective at the receptor level, as is the case with L1210 cells (Aguet 1980). However, several other resistant human and mouse cell lines have been developed which do have receptors, and these have been used to investigate the role of the two induced enzymes whose activities are activated by dsRNA, (namely, the protein kinase and 2-5A synthetase) in interferon-induced growth inhibition. These studies have shown that with one exception, (where the enzyme is constitutive) the 2-5A system is induced in the resistant lines. Levels of the oligonucleotides, however, have only been looked at in one case, the sensitive and resistant Daudi cells, and in both cell types the levels were too low for detection

(Silverman, Watling, Balkwill, Trowsdale, and Kerr 1982). The protein kinase is also induced in those resistant cell lines which have been tested. Thus the results from work on interferon-induced proteins in cell mutants resistant to interferon's growth inhibitory effect do not point to a consistent role for the 2-5A synthetase or the protein kinase in interferon-induced growth inhibition, although some aspects of the 2-5A system may be involved in some cell types. It has been suggested that the interferon-induced dsRNA dependent kinase may phosphorylate ornithine decarboxylase (ODC) and thus inactivate it, in the same way that a polyamine-dependent kinase regulates the activity of ODC in the slime mould *Rhysarum polycephalum* (Sekar, Atmar, Krim, and Kuehn 1982). This is an appealing idea which has not as yet received experimental confirmation.

It is interesting to note that undifferentiated embryonal carcinoma cells have been shown to be insensitive to either the antiviral or anti-growth effects of interferon (Burke, Graham, and Lehman 1978). In this case, receptors are present and 2-5A synthetase is induced, although the protein kinase appears not to be (Wood and Hovanessian 1979). Since the responsive differentiated cells do show inducible kinase activity, there is a possibility that the kinase is involved, although again it may just reflect a block in a controlling event in the interferon response.

5.5.2. Studies on cell lines with defective functions

Lines defective in cAMP-dependent kinase

Schneck, Rager-Zisman, Rosen, and Bloom (1982) have developed a series of macrophage cell lines which, unlike the parent line, have defects in the cyclic AMP-dependent kinase. Their results provide evidence for a central role for cAMP in interferon-induced growth inhibition in these cell lines. In the parent cell line, cAMP has been shown to inhibit cell growth, while the mutant cell lines with a defective kinase show a much reduced sensitivity to cAMP and to interferon's activity both as an inhibitor of cell growth and as an enhancer of cell function. These results are in contrast to those reported by Bannerjee, Baksi, and Gottesman (1983) using another series of mutant Chinese hamster ovary (CHO) cell lines. These mutant cells, which are also defective in the cAMP-dependent protein kinase, show no reduction in sensitivity to interferon as an inhibitor of cell growth. One must conclude from the studies relating to cAMP and interferon that the nucleotide may have a role to play in mediating the inhibition of cell growth induced by interferon in some cell types, but it cannot be considered to be a general second messenger in all systems. Indeed, cAMP is a mitogenic signal for many cell types, and in these cells, the nucleotide plays no direct or indirect role in interferon's action as a growth inhibitor (Ebsworth, Taylor-Papadimitriou, and Rozengurt 1984). It should also be mentioned that a role for cyclic GMP has been proposed in interferon action in inhibiting cell growth, but studies with resistant cell lines have not completely supported this idea (Tovey 1982).

Lines with a defective thymidine kinase

It has been suggested that inhibition of cell growth by interferon depends on a functioning thymidine kinase (TK). Interferons have been shown to inhibit uptake of thymidine in a variety of systems, and where this is rate limiting, as it is in media with low folic acid levels, such an inhibition could conceivably lead to an inhibition of cell growth. However, the correlation of interferon inhibition of cell growth with a functioning thymidine kinase, originally seen in L929 cells, could not be shown with a series of S49T lymphoma mutants (for review see Taylor-Papadimitriou *et al.* 1984). It should be pointed out that for these studies the medium in which cells are grown is critical; it might be interesting to determine if interferon inhibits growth more profoundly in medium 199 (which contains low levels of folic acid) and if this is reversed by exogenous thymidine. It is also possible that modulation of thymidine incorporation may be a result of interferon action in inhibiting cell growth rather than a primary cause of this inhibition.

The results with interferon-resistant and mutant cell lines have this far been inconclusive, but only a limited number of functions have been analyzed. Membrane changes and the importance of cytoskeleton components for example have not been examined, and this approach must still be considered a logical one worth pursuing.

5.5.3. Interferon and the mitogenic signal

Growth regulation in cultured cells

The use of cell culture as a tool for the study of a variety of fundamental biological problems is widely accepted. In particular, it is recognized that cells in culture provide an experimental system for studying the factors that modulate cell proliferation and differentiation without many of the complexities of whole animal experimentation. Such *in vitro* studies have shown that cells proliferate in response to certain growth factors, some of which may be found in the serum usually used to supplement tissue culture media. Many normal cells cease to proliferate and become quiescent with a G1 DNA content when they have depleted the serum of its growth-promoting activity, and such 'quiescent' cells can be re-stimulated to enter DNA synthesis by the addition of fresh serum, or by a combination of purified growth factors added to serum-free medium (Rozengurt 1983, and see Fig. 5.2). Such model systems in general, and the Swiss 3T3 cell line in particular, are important for studying the molecular events which occur in cells moving from quiescence to growth, since the cells are synchronized. Before discussing the effects of interferons on cell growth it is appropriate to consider briefly what is known of the signals generated by the mitogens in such systems.

Considerable progress has been made in the study of the growth factors

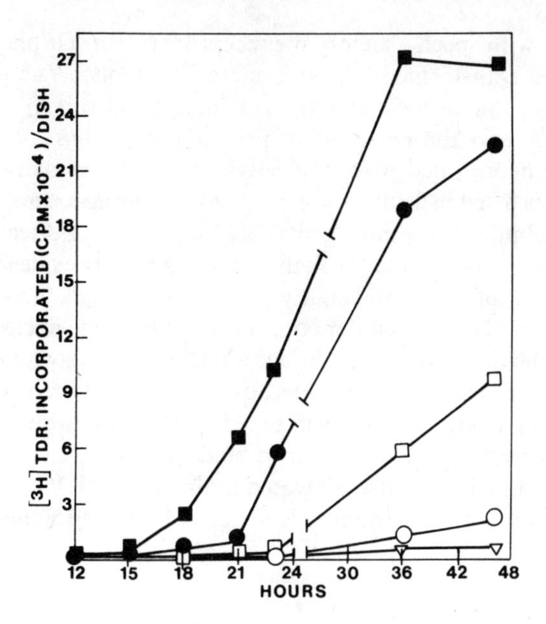

Fig. 5.2. Entry of quiescent Swiss 3T3 cells into DNA synthesis after stimulation with serum (■—■, □—□) or growth factors (●—●, ○—○) in the presence (□—□, ○—○) or absence (■—■, ●—●) of mouse L cell interferon (1000 i.u./ml). ○—○ represents unstimulated cells.

themselves, their interactions with receptors and the initial signals generated by this interaction (Rozengurt 1983; Rozengurt and Collins 1983). Table 5.5 lists the various growth factors which have been found to act on Swiss 3T3

Table 5.5. *Chemically diverse factors which stimulate DNA synthesis in Swiss 3T3 cells*

Polypeptide factors	EGF, PDGF, IGFs[a]
Neurohypophyseal hormones	Vasopressin, oxytocin, analogues
Tumour promoters	Phorbol esters, (TPA), teleocidin
Regulatory peptides	Bombesin
Vitamin A derivatives	Retinoic acid
Permeability modulators	Melittin
Cyclic nucleotide elevating agents	Cholera toxin, adenosine agonists, cAMP derivatives
Microtubule disrupting agents	Colchicine, colcemid, vinblastine, podophyllotoxin, nocodazole
Polypeptide released by transforming cells	FDGF, TGFs[b]

[a]epidermal growth factor, platelet-derived growth factor, insulin-like growth factors.
[b]Fibroblast-derived growth factor, transforming growth factors.

cells. Some of these, such as insulin, EGF, PDGF, vasopressin and bombesin are naturally produced polypeptides which are known to operate *in vivo,* and which interact with specific membrane receptors (often self-phosphorylating kinases). Some transformed and tumour cells produce factors which are structurally and functionally similar (and may be identical) to the normal growth factors, and the constitutive production of these factors may be related to the uncontrolled growth of cells seen in malignant disease. Some of the other agents listed in Table 5.5 (e.g. cAMP-elevating agents, permeability modulators, tubulin disrupting agents) act by generating directly one of the signals elicited by the normal growth factors. There is evidence suggesting that the opening of ion permeability pathways through the plasma membrane, changes in the intracellular concentration of cyclic nucleotides and/or alterations in the organization of the cytoskeleton may play a role as internal signals in the regulation of the proliferative response (Rozengurt 1983), and this is illustrated diagrammatically in Fig. 5.3. The tumour promoters form a special class of reagents which appear to bind specifically to protein kinase C (Nishizuka 1983), which is also activated indirectly by PDGF and FDGF in intact fibroblastic cells (Rozengurt, Rodriguez-Pera, and Smith 1983).

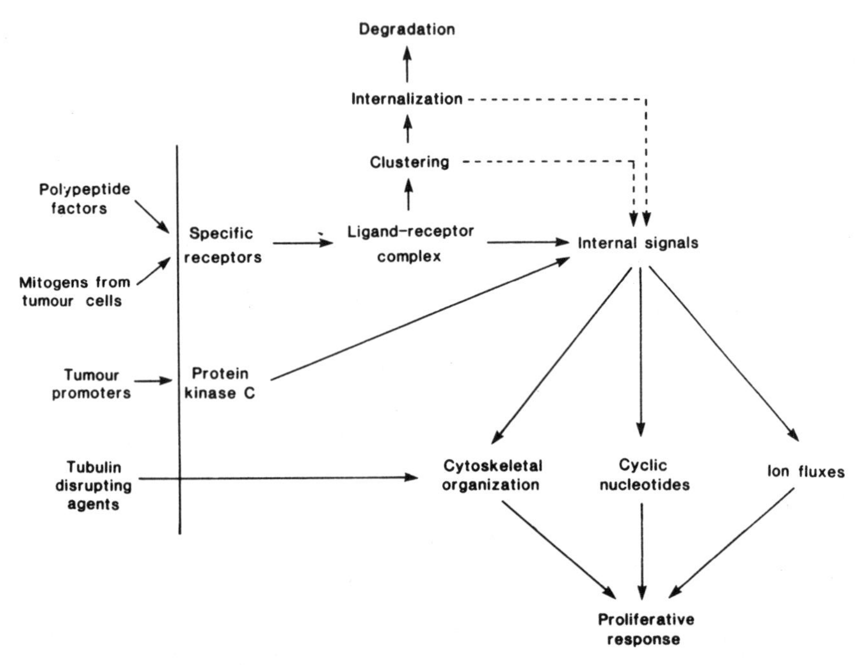

Fig. 5.3. Possible internal signals generated by various growth factors in Swiss 3T3 cells.

In contrast to the progress made in understanding some of the early events following growth factor stimulation, there is little definite information available regarding the latter events which must occur in G1 to lead up to the onset of DNA synthesis. As can be seen in Fig. 5.2, 12 h may elapse before DNA synthesis begins, and although it is clear that a general increase in size, RNA synthesis etc. need to occur, specific reactions have not been identified which are crucial to DNA synthesis. It has been hypothesized for a long time that labile proteins are required for cells to progress through G1 (Schneiderman, Dewey, and Highfield 1971). Recent studies have extended this hypothesis and attempted to characterize the new gene expression messages (Cochran, Reffel, and Stiles 1983) and labile proteins (Croy and Pardee 1983), which are being made.

Inhibition by interferon of the mitogenic signal in synchronized cells.

The effect of interferons on growth have been examined using either unsynchronized cycling cells or synchronized cells moving from quiescence to growth after stimulation by growth factors. Most of these studies show that G1 and G2 can be extended, and whether G1 is extended more than G2 depends on the cell type (for review see Taylor-Papadimitriou 1983). As can be seen from Fig. 5.2, G1 is extended by interferon treatment of quiescent Swiss 3T3 cells stimulated to proliferate with either serum or combinations of growth factors. These cells therefore provide a useful model system for asking questions about the molecular basis for the inhibition by interferon of entry into DNA synthesis.

As indicated earlier, polypeptide growth factors initiate their action in quiescent 3T3 cells by binding to receptors and generating various intracellular signals which act synergistically to induce reinitiation of DNA synthesis and cell division. Where binding of the growth factor can be measured using isotopically labelled ligand, as in the case of EGF, interferon can be shown to have no effect on such binding (Lin, Ts'o, and Hollenberg 1980). Similarly, neither cAMP levels nor early ion fluxes in both quiescent and stimulated cells are affected by interferon (Ebsworth *et al.* 1984; Taylor-Papadimitriou *et al.* 1984). It would appear, therefore, that in Swiss 3T3 cells, binding of growth factors and the initial signals generated by the receptor interaction are not blocked by interferon.

One observation made in these cells suggests that interferons may inhibit a common step after the convergence of the separate signals which may be generated by the different growth factors. Thus the extent of inhibition of entry into DNA synthesis is inversely proportional to the intensity of the mitogenic signal, a strong inhibitory effect being seen in cells stimulated with only two growth factors, and no inhibition at all being evident in cells stimulated with five factors (Taylor-Papadimitriou, Shearer, and Rozengart 1981). A candidate for such a common interferon-sensitive step is the reaction catalyzed by ornithine decarboxylase, which is the rate limiting and first step in the production of the polyamines. ODC activity is increased in all proliferating cells and this increase is dramatically inhibited in interferon-treated cells (Sreevalsan, Rozengurt, Taylor-Papadimitriou, and Burchell

1980). Moreover, it is found that, while the inhibition of ODC induction by interferon is seen when several mitogens are used to stimulate 3T3 cells, the absolute level in the highly stimulated interferon-treated cell is much higher than the level in the comparable cell stimulated with fewer mitogens. It could therefore be argued that the levels of ODC are regulating the rate of entry into S.

While the inhibition of the induction of ODC activity may be crucial to a sustained effect of interferon on cell growth, it does not appear to be the only interferon-sensitive step in cells moving through G1. The peak of ODC activity is seen 4–6 h after stimulating 3T3 cells with growth factors, and it is possible to add interferon after this point and still obtain more than 50 per cent inhibition of DNA synthesis. This indicates that some event (events) in the second half of G1 which is (are) crucial to DNA synthesis is also inhibited by interferon. One clue as to what such an event might be comes from the observation that tubulin-disrupting agents (such as colchicine) can be added several hours after other mitogens and still stimulate the rate of entry into DNA synthesis. Moreover, no effect is seen on entry into S if the tubulin-disrupting agent is added with the other growth factors and removed before 4 h (Wang and Fozengurt 1983). It is quite possible therefore that interferon has the opposite effect to tubulin-disrupting agents and stabilizes the tubulin network and by so doing inhibits entry into DNA synthesis even when added late in G1. Such a suggestion is supported by previous work showing that interferon action on cell and tumour growth is enhanced by agents which stabilize tubulin (Bourgeaude and Chany 1979). Another effect of interferon which implicates the tubulin network is the increase in the level of tubulin mRNA noted in interferon-treated cells (Fellous *et al.* 1982).

5.6. INTERFERON–RECEPTOR INTERACTIONS

Since purified preparations of single species of interferons have become available, it has been possible to follow binding to cells using radiolabelled ligand. Such studies, which have been done mainly with mouse β interferon and with HuIFN-α, indicate that there are specific receptors for interferons in the cell membrane, although the number is very small, being of the order of < 1000 even for those cells with the highest receptor levels. Competition studies suggest that on homologous cells, the interferons of the α and β class share a common receptor, but a separate receptor is used by the corresponding interferon of the γ class. It is relevant to ask whether the initial signals generated by the interaction of an interferon with the membrane receptor are different in cells which are sensitive and cells which are resistant to the growth inhibitory effect.

Until the human interferons were studied in detail, interferons were considered to be species specific. Human interferons of the α-class, however, were found to induce an antiviral state in other species, bovine and feline cells being particularly sensitive. Bovine cells (usually the MDBK cell line) are widely used to assay HuIFNs of the α-class since all the different subclasses give similar specific activity when titrated for antiviral activity on these cells. This is in contrast to the situation on human cells where HuIFN-α_1 is 100 times less effective as an antiviral agent than HuIFN-α_2 (Streuli, Hall, Boll, Nagata, and Weissman 1981). In spite of being highly sensitive to the antiviral effect of HuIFN-α, we have found bovine cells to be resistant to the antigrowth effect of this interferon (Taylor-Papadimitriou, Shearer, Balkwill, and Fantes 1982). Since all the bovine cells tested (some of which were very sensitive to the antiviral effect) were resistant to the antigrowth effect and most human cells (even those with low sensitivity to the antiviral effect) showed some degree of growth inhibition, it seems likely that it is not the response of the cell which is different in the two species but rather the signal(s) generated by the membrane interaction of the HuIFN-α. We have based our experimental approach on this reasoning and have examined in detail the kinetics of binding of ^{125}I-HuIFN-α_2 at 4°C to a range of human and bovine cells.

Previous data studying the antiviral action of human interferons and also their binding to bovine and human cells suggested that the interaction of HuIFN-α with the bovine and human membrane receptor is not identical. Thus, HuIFN-β has no antiviral action on bovine cells and cannot compete with HuIFN-α for binding to these cells (Yonehara, Yonehara-Takahashi, and Ishii 1983*a*) although HuIFN-α and HuIFN-β appear to compete for the same receptor on human cells. Furthermore, the different antiviral activity of HuIFN-α_1 (and various α hybrids) on human and bovine cells has been shown to be due to its affinity for the receptor (Yonehara, Yonehara-Takahashi, Ishii, and Nagata 1983*b*). Finally, a point which is not always considered is that binding data with bovine cells can be and has been obtained at 4°C (Arnheither, Ohno, Smith, Gutte, and Zoon 1983) while most of the work with human is done at 37° or 21°C and usually with the highly sensitive Daudi cell (Mogensen and Bandu 1983; Hannigen, Gewert, Fish, Read, and Williams 1983; Branca and Baglioni 1981). We have made a comparison of the kinetics of binding of ^{125}I-HuIFN-α_2 to a range of bovine and human cell lines at 4°C where the effector functions of the ligand–receptor complex cannot proceed (Taylor-Papadimitriou and Shearer 1984). Our results clearly show that the kinetics of binding the HuIFN-α_2 to bovine cells follow the simple kinetics predicted from the binding of a single molecular species of ligand to a single high affinity membrane receptor. The dissociation constant for HuIFN-α_2 was found to be 3×10^{-11} for the three bovine cell types. This figure is similar to the K_d reported for MDBK cells (Yonehara *et al.* 1983*b*).

In sharp contrast to the results with bovine cells, it was found that the kinetics of binding of HuIFN-α_2 to human cells are complex. At concentrations of interferon less than 100–200 i.u., receptor occupancy appears to increase the rate of binding of HuIFN-α_2. This is seen clearly with two relatively insensitive cell lines (T47D and ICRF-23), and after disruption of the tubulin network, with a very sensitive cell line (BT20). The fact that colchicine can affect the binding curve of HuIFN-α_2 for BT20 cells suggests that receptors bound to an organized tubulin network have a different affinity for HuIFN-α_2 than those not bound. On the other hand, disruption of the tubulin network in bovine cells does not affect the dose binding curve. The fact that there are a relatively high number of receptors in BT20 cells, and that at least some of these interact with the tubulin network may explain why the interferon–receptor interaction in these cells is effective in inhibiting cell growth even at low concentrations.

While it is not possible to draw definite conclusions from the kinetic data about the molecular basis for the complex kinetics seen with human cells, they suggest that formation of interferon dimers on the receptor could be involved. Stabilization of binding of one monomer of HuIFN-α_2 by binding of a second could explain the positive co-operative binding we see with human cells [as is seen with the binding of the λ repressor to DNA which shows similar kinetics (Johnson, Poteete, Lauer, Sauer, Acjers, and Ptashne 1981)]. HuIFN-α molecules show a pH and concentration-dependent reversible aggregation, forming dimers and even higher oligomers, and this may reflect a functional role for dimer formation on the receptor (Pestka, Kelder, Familletti, Moshera, Crowl, and Kempner 1981). Formation of interferon dimers from monomers bound to separate receptors might occur where receptors are more abundant and transposable through being bound to structural elements (e.g. as in BT20 cells).

The possibility that more than one α monomer may bind to the receptor is interesting when we consider that the α and β interferons appear to share a common receptor on human cells and that HuIFN-β normally exists as a dimer (Pestka *et al.* 1983). If only monomers can bind to the bovine receptor, it is not surprising that HuIFN-β is not bound. The binding of two different α molecules to one receptor in human cells would also allow for a range of ligand interactions with the mixtures of HuIFN-α normally naturally produced by lymphocytes. Synergy has already been reported for α_1 and α_2 and this could be operating at the receptor level (Orchansky, Gosen, and Rubinstein 1983).

5.7. CONCLUSIONS

The studies of interferon's effects on cell growth and differentiation can be considered at different levels of understanding. Firstly, it is very clear that highly purified preparations of IFN or those produced by recombinant DNA technology produce a potent inhibition of cell growth and modulate cell function in many different cell types including normal and maligant cells.

However, the precise sensitivity of the cells to interferon is a complex parameter, an estimate of which requires the use of chemically defined medium to grow the cells and thereby standardize the conditions of the assay. Secondly, if the action of interferon on cells is considered at the subcellular level it can again be said with certainty that profound changes in cellular structure and in cell metabolism are seen in the interferon-treated cells. Here it must be remembered that these changes may vary from one cell type to another, as do the structures and metabolic profiles affected, although some effects, such as inhibition of ODC activity, appear to be common to many cell types and to be induced by all classes of interferons.

In contrast to the definitive statements which can now be made regarding the ability of interferons to inhibit cell growth and regulate cell behaviour and metabolism, it is not yet possible to draw definite conclusions regarding the mechanisms underlying these effects. This is hardly surprising, since growth and differentiation are highly complex phenomena, only now being defined at the molecular level, and interferons induce manifold changes in cells. The task therefore of determining which of the interferon-induced changes are relevant to its action, for example, as an inhibitor of cell growth, is a difficult one although some clues may be appearing. The use of cell culture (particularly synchronized systems) for studying growth regulation has led to the identification of both a growth-related enzyme (ODC) which is inhibited by all three types of interferon, and a cytoskeletal component (tubulin) which appears to be implicated in interferon induced growth inhibition: these findings might provide a starting point for investigations into the mechanisms underlying this inhibition.

Attempts to define cell responses crucial to interferon-induced growth inhibition using resistant and sensitive cells have not yet yielded consistent positive correlations, although it seems clear that the major interferon-induced proteins are not crucially involved. This approach however seems to us to be one which should still be pursued, although the investigation of the interferon–cell interaction should probably start at the receptor level, looking for initial signals which may be found only in the sensitive cells. Indeed, the major gap in our understanding in the mechanism of action of interferon is to define the molecular events whereby receptor occupancy by this ligand elicits multiple biological responses in the cell.

5.8. REFERENCES

Aguet, M. (1980). High-affinity binding of ^{125}I-labelled mouse interferon to a specific cell surface receptor. *Nature* **284,** 459–61.

Ankel, H., Krishnamurti, C., Besancon, F., Stefanos, S., and Falcoff, E. (1980). Mouse fibroblast (type I) and immune (type II) interferons: pronounced differences in affinity for gangliosides and in antiviral and antigrowth effects on the mouse leukemia L-1210R cells. *Proc. Nat. Aad. Sci. USA* **77,** 2528–32.

Arnheiter, H., Ohno, M., Smith, M. R., Gutte, B., and Zoon, K. C. (1983).

Orientation of a human leukocyte interferon molecule on its cell surface receptor: carboxy-terminus remains accessible to a monoclonal antibody made against a synthetic interferon peptide. *Proc. Nat. Acad. Sci. USA* **80,** 2539–43.

Banerjee, D. K., Baksi, K., and Gottesman, M. M. (1983). Genetic evidence that action of cAMP-dependent protein kinase is not an obligatory step for antiviral and antiproliferative effects of human interferon in Chinese hamster cells. *Virology* **129,** 230–8.

Bourgeaude, M. F., and Chany, C. (1979). Effect of sodium butyrate on the antiviral and anti-cellular action of interferon on normal and MSV-transformed cells. *Int. J. Cancer.* **24,** 314–18.

—— Rousset, S., Paulin, D., and Chany, C. (1981). Reorganization of the cytoskeleton by interferon in MSV-transformed cells. *J. Interferon Res.* **1,** 323–32.

Branca, A. A. and Baglioni, C. (1981). Evidence that types I and II interferons have different receptors. *Nature* **294,** 768–70.

Burke, D. C., Graham, C. F., and Lehman, J. M. (1978). Appearance of interferon inducibility and sensitivity during differentiation of murine teratocarcinoma cells *in vitro*. *Cell* **13,** 243–8.

Chany, M. C., and Vignal, M. (1968). Etude du mechanisme de l'état refractaire des cellules à la production d'interferon, après inductions repetées. *C.R. Hebd. Séarc. Acad. Sci. Paris,* **267,** 1798–800.

Cochran, B. H., Reffel, A. C., and Stiles, C. D. (1983). Molecular cloning of gene sequences regulated by platelet-derived growth factor. *Cell* **33,** 939–47.

Croy, R. G., and Pardee, A. B. (1983). Enhanced synthesis and stabilization of Mr 68 000 protein in transformed BALB/c-3T3 cells: candidate for restriction point control of cell growth. *Proc. Nat. Acad. Sci. USA* **80,** 4699–703.

Ebsworth, N., Taylor-Papadimitriou, J., and Rozengurt, E. (1984). Cyclic AMP does not mediate inhibition of DNA synthesis by interferon in mouse Swiss 3T3 cells. *J. Cell. Physiol.* **120,** 146–150.

Fantes, K. H. and Allen, G. (1981). Specific activity of pure human interferons and a non-biological method for estimating the purity of highly purified interferon preparations. *J. Interferon Res.* **1,** 465–74.

Fellous, A., Ginsberg, I., and Littauer, U. Z. (1982). Modulation of tubulin mRNA levels by interferon in human lymphoblastoid cells. *EMBO J.* **7,** 835–9.

Fleischman, W. R., Kleyn, K. M., and Bron, S. (1980). Potentiation of antitumour effect of virus induced interferon by mouse immune interferon. *J. Nat. Cancer Inst.* **65,** 963–6.

Gresser, I. and Tovey, M. G., (1978). Antitumour effects of interferon. *Biochim. Biophys. Acta* **516,** 231–47.

—— Maury, C., and Brouty-Boye, D. (1972). Mechanism of antitumour effect of interferon in mice. *Nature* **239,** 167–8.

Hannigan, G. E., Gewert, D. R., Fish, E. N., Read, S. E., and Williams, B. R. G. (1983). Differential binding of human interferon-α subtypes to receptors on lymphoblastoid cells. *Biochem. Biophys. Res. Commun.* **110,** 537–44.

Johnson, A. D., Poteete, A. R., Lauer, G., Sauer, R. T., Ackers, G. K., and Ptashne, M. (1981). λ Repressor and Cro-components of an efficient molecular switch. *Nature* **294,** 217–23.

Keay, S. and Grossberg, S. E. (1980). Interferon inhibits the conversion of 3T3-L1 mouse fibroblasts into adipocytes. *Proc. Nat. Acad. Sci. USA* **77,** 4099–103.

Lin, S. L., Ts'o, P. O. P., and Hollenberg, M. D. (1980). The effects of interferon on epidermal growth factor action. *Biochem. Biophys. Res. Commun.* **96,** 163–74.

McMahon, M. and Kerr, I. M. (1983). The biochemistry of the antiviral state. In *Interferons: from molecular biology to clinical application. SGM Symposium 35* (ed. D. C. Burke and A. G. Morris). Cambridge University Press, Cambridge.

Mogensen, K. E. and Bandu, M.-T. (1983). Kinetic evidence for an activation step following binding of human interferon-α_2 to the membrane receptors of Daudi cells. *Eur. J. Biochem.* **134,** 355–64.
Nishizuka, Y. (1983). Phospholipid degradation and signal translation for protein phosphorylation. *Trends Biochem. Sci.* January, 13–16.
Orchansky, P., Gosen, T., and Rubinstein, M., (1983). Isolated subtypes of human interferon α: synergistic action and affinity for the receptor. In *The biology of the interferon system,* (ed. E. De Maeyer and H. Schellekens, pp. 183–8. Elsevier, Amsterdam.
Paucker, K., Cantell, K., and Henle, W. (1962). Quantitative studies on viral interference in suspended L-cells. III. Effect of interfering viruses and interferon on the growth rate of cells. *Virology* **17,** 324–34.
Pestka, S., Kelder, B., Familletti, P. C., Moshera, J. A., Crowl, R., and Kempner, E. S. (1983). Molecular weight of the functional unit of human leucocyte, fibroblast, and immune interferons. *J. Biol. Chem.* **258,** 9706–9.
Pfeffer, L. M., Landsbergery, F. R., and Tamm, I. (1981). Beta interferon induced time-dependent changes in the plasma membrane lipid-bilayer of cultured cells. *J. Interferon Res.* **1,** 613–19.
Rozengurt, E. (1983). Growth factors, cell proliferation and cancer: an overview. *Mol. Biol. Med.* **1,** 169–81.
—— and Collins, M. (1983). Molecular aspects of growth factor action: receptors and intracellular signals. *J. Pathol.* **141,** 309–31.
—— Rodriguez-Pena, M., and Smith, K. A. (1983). Phorbol esters, phospholipase C, and growth factors rapidly stimulate the phosphorylation of a Mr 80 000 protein in intact quiescent 3T3 cells. *Proc. Nat. Acad. Sci. USA* **80,** 7244–8.
Schneck, J., Rager-Zisman, B., Rosen, O. M., and Bloom, B. R. (1982). Genetic analysis of the role of cAMP in mediating effects of interferon. *Proc. Nat. Acad. Sci. USA.* **79,** 1879–83.
Schneiderman, M. H., Dewey, W. C., and Highfield, D. P. (1971). Inhibition of DNA synthesis in synchronized Chinese hamster cells treated in G1 with cycloheximide. *Exp. Cell Res.* **67,** 147–55.
Sekar, V., Atmar, V. J., Krim, M. and Kuehn, G. C. (1982). Interferon induction of polyamine dependent protein kinase activity in Ehrlich ascites tumour cells. *Biochem. Biophys. Res. Commun.* **106,** 305–11.
Silverman, R. H., Watling, D., Balkwill, F. R., Trowsdale, J., and Kerr, I. M. (1982). The ppp(A2'p)$_n$A and protein kinase systems in wild-type and interferon-resistant Daudi cells. *Eur. J. Biochem.* **126,** 333–41.
Sreevalsan, T., Rozengurt, E., Taylor-Papadimitriou, J., and Burchell, J. (1980). Differential effect of interferon on DNA synthesis, 2-deoxyglucose uptake and ornithine decarboxylase activity in 3T3 cells stimulated by polypeptide growth factors and tumour promoters. *J. Cell. Physiol.* **104,** 1–9.
Streuli, M., Hall, A., Boll, W., Stewart, II, W., Nagata, S., and Weissmann, C. (1981). Target cell specificity of two species of human interferon-α produced in *E. coli* and hybrid molecules derived from them. *Proc. Nat. Acad. Sci. USA* **78,** 2848–52.
Taylor, J. (1964). Inhibition of interferon action by actinomycin. *Biochem. Biophys. Res. Commun.* **14,** 447–51.
Taylor-Papadimitriou, J. (1980). Effects of interferon on cell growth and function. In *Interferon 2,* (ed. I. Gresser) pp. 13–36. Academic Press, London.
—— (1983). The effects of interferon on the growth and function of normal and malignant cells. In *Interferons: from molecular biology to clinical application. SGM Symposium 35.* (ed. D. C. Burke and A. G. Morris) pp. 109–47. Cambridge University Press, Cambridge.
—— Shearer, M., and Rozengurt, E. (1981). Inhibitory effect of interferon on cellular

DNA synthesis: modulation by pure mitogenic factors. *J. Interferon Res.* **1,** 401–9.
—— —— Balkwill, F. R., and Fantes, K. H. (1982). Effects of HuIFN-α_2 and HuIFN-α (Namalwa) on breast cancer cells grown in culture and as xenografts in the nude mouse. *J. Interferon Res.* **2,** 479–91.
—— Ebsworth, N., and Rozengurt E. (1984). Possible mechanisms of interferon-induced growth inhibition. In *Mediators in cell growth and differentiation* (ed. R. J. Ford and A. L. Maizel) pp. 283–98. Raven Press, New York.
Tovey, M. G. (1982). Interferon and cyclic nucleotides. *Interferon* **4,** 23–46.
Wang, Z. W. and Rozengurt, E. (1983). Interplay of cyclic AMP and microtubules in modulating the initiation of DNA synthesis in 3T3 cells. *J. Cell Biol.* **96,** 1743–50.
Wang, E., Pfeffer, L. M., and Tamm, I. (1981). Interferon increases the abundance of submembraneous microfilaments in HeLa-S3 cells in suspension culture. *Proc. Nat. Acad. Sci. USA* **78,** 6281–5.
Wood, J. N. and Hovanessian, A. G. (1979). Interferon enhances 2-5A synthetase in embryonal carcinoma cells. *Nature* **282,** 74–6.
Yonehara, S., Yonehara-Takahashi, M., and Ishii, A. (1983*a*). Binding of human interferon-α to cells of different sensitivities: studies with internally radiolabeled interferon retaining full biological activity. *J. Virol.* **45,** 1168–71.
——, —— Ishii, A., and Nagata, S. (1983*b*). Different binding of human interferon α_1 and α_2 to common receptors on human and bovine cells. *J. Biol. Chem.* **258,** 9046–9.

6 Interferons and infectious disease

G. M. Scott

6.1. INTRODUCTION

Interferon was discovered because one of its properties, perhaps the most important one, had to do with the inhibition of virus replication in tissue culture. But after over 25 years of intensive investigation by many workers and several thousand reports on all aspects of the interferon system, it is still not clear whether interferon is actually necessary for recovery from acute or chronic virus infection in humans. This is because definitive experiments using an interferon antibody during an experimental acute virus infection in volunteers are not ethical because of the potential risk. In order to investigate the role of interferon in humans, we can examine the production of interferon under different circumstances, the effects of endogenous or exogenous interferons and search for disease states where there appears to be a failure of interferon production. In animals we may go one step further and neutralize the effects of endogenous interferon. The inference from what we know about the production of interferon in acute infections, about the antiviral activity of viral-induced interferon and its effects on the immune system and experiments with model infections in animals, is that it is a crucial part of the innate response to virus invasion and replication. The production of immune interferon by immunocompetent T-lymphocytes in response to antigenic stimuli is a reflection of the learned response and so may be equally important in the anamnestic response to recurrent infection or in prolonged acute or chronic disease.

No major qualitative differences in the biological properties of different human interferons have been described except for different affinities for cells of diverse species. There are, however, some quantitative differences, for example between viral (α or β, type I) and immune (γ, type II) interferons in the enhancement of expression of certain HLA antigens, but as interferons are measured by comparing their antiviral or immunoreactive behaviour against standards in cells lines or in immunoassays, differences in the magnitude of *in vitro* responses may be meaningless. At present, we do not know enough about these differences *in vivo* to be able to categorize responses to certain infections as primarily 'type I' or primarily 'type II', and then to draw different conclusions about the pathophysiology of these infections. If all interferons have the same properties to a greater or lesser degree, their production in innate and learned responses suggests that their role extends beyond direct restriction of local virus replication.

The naïve interpretation of the observation that interferon prevents virus replication in cells *in vitro* is that this is the way in which virus replication is held in check locally in the intact organism until specific humoral and cell-mediated mechanisms can eradicate the infection. It is more likely that interferon is just as important in initiation of these latter mechanisms through a variety of effects on the immune system, as discussed in detail in Chapter 4. However, the effects of interferons on the whole organism are complicated by their ability to stimulate a variety of seemingly unrelated pathophysiological systems such as prostaglandins and corticosteroids as well as a host of important intracellular pathways (Chapters 3–5) which have their own diverse effects on immune functions.

That interferons can be induced *in vitro* by a range of bacteria and bacterial products and protozoa, and also found in bacterial and parasitic infections *in vivo,* suggest a broader role in infections than has been considered. The findings that exogenous interferons cause certain 'unwanted' effects in animals and man and the presence of circulating interferons in various autoimmune diseases suggest a possible role in symptomatology and even pathogenesis of disease.

Recovery from acute virus infection involves a complex variety of host-defence mechanisms. It has emerged that interferon may affect virtually every facet of host defence including non-specific inflammation, antigen expression, phagocytosis, antibody formation and cell-mediated resistance. Although it is not possible to make definitive or didactic statements about the role of natural interferon in host response to infection, in this brief review we can consider some of the circumstantial evidence collected so far.

6.2. INTERFERON PRODUCTION IN HEALTHY INDIVIDUALS

Interferon cannot usually be detected in the serum of healthy individuals. Assays of interferon in serum are complicated because non-specific factors may enhance or suppress virus replication. However, bovine cells or trisomic (Chr 21) human cells are highly sensitive to certain human interferon subspecies and new immunoassays using monoclonal antibodies comfortably allow detection of $\leqslant 10$ i.u. ml^{-1}; some laboratories claim far better sensitivity than that ($\leqslant 1$ i.u. ml^{-1}). With these assays, some workers claim (and the claim is controversial because interferon assays are notoriously imprecise) that there is a background antiviral activity in normal serum of around 1–4 i.u. ml^{-1}. There is a further question, however, whether interferon is produced under normal physiological conditions and this has been proposed by Bocci (1981). The arguments are that interferon-induced enzymes may be found in serum and peripheral white cells of healthy individuals and that certain circulating lymphocytes, perhaps natural killer cells, have the capacity to make interferon spontaneously *in vitro*. Anyway, the levels of interferon and interferon-induced enzymes are very low compared with those found in disease. Interferon is cleared very rapidly from the serum, probably absorbed from the proximal renal tubule and degraded there, but Bocci, Muscettola, Paulesu, and Grasso (1984) have recently shown the presence of moderate levels of antiviral activity in lymph from apparently healthy rabbits. This had some of the characteristics of viral-induced interferon and some of immune interferon, so at present we must speculate that a mixture of interferons may be produced spontaneously by the gut-associated lymphoid tissue. No cyclical variation was demonstrable over a 4-h period although some nyctohemeral variation would be expected and deserves further investigation. If interferon is produced spontaneously, does it have a role? The gut-associated lymphoid tissue is continuously challenged with old and new antigens absorbed from the gut and it is tempting to speculate that because interferon has time- and dose-dependent opposing effects on immunological functions *in vitro,* it has something to do with maintaining immunological homeostasis.

6.3. INTERFERON IN DISEASE STATES

6.3.1. Measurement of the interferon response

Several ways of assessing the role of the interferon system in recovery from disease may be considered. First, levels of interferon antiviral activity may be measured in various body fluids including serum, the contents of vesicles of some common viral exanthems and tissue biopsy extracts. A temporal relationship between interferon detection and recovery from virus infection may be demonstrated. If interferon is not detectable, peripheral white cells or tissue cells may be shown to be actively secreting interferon by immunofluorescence using monoclonal antibodies, but this method has not yet been described in disease. Secondly, and perhaps a more appropriate parameter, the enzymes induced by interferon may be measured in serum and peripheral blood mononuclear cells. At present, these assays are complex and time consuming, so are unsuitable for routine diagnostic laboratories. Moreover the specificity of the enzyme response to interferon and not to other physiological substances (such as thyroxine) needs clarification.

Thirdly, peripheral blood mononuclear cells may be infected with a virus to determine whether they are in an 'antiviral state' or capable of sustaining viral replication; and fourthly, peripheral mononuclear cells may also be stimulated with powerful interferon inducers (such as paramyxovirus, poly rI. poly rC or mitogen) to determine their ability to produce interferon under optimal conditions *in vitro*. Finally, tests of interferon-linked functions such as natural killer cell activity may also indirectly reflect the activity of the interferon system.

6.3.2. Deficiency of interferon and progression of viral disease

Using these tests we can search for diseases where there appears to be a relative or absolute deficiency of interferon production and relate them to the pathology of the diseases themselves or to an alteration in response to virus infection (see Table 6.1). Such a comprehensive approach has been made by

Table 6.1. *Examples of apparent deficiencies of interferon (IFN) production and their possible origins*

	Determined by		
	Virus	Disease	Host
Deficiency in synthesis of:			
IFN *in vivo*	RSV (nose, serum); acute hepatitis B (serum); fatal influenza (lung)	Hodgkin's disease (herpes zoster)	
IFN-α *in vitro*		Acute leukaemia; Hodgkin's disease	Young age, marasmus
	Various fulminant infections: hepatitis or encephalitis		
IFN-γ *in vitro*		Multiple sclerosis	Syndrome of susceptibility to chronic EBV infection
		Leprosy and tuberculosis; acquired immune deficiency	
2'5'-A synthetase	Various fulminant infections: hepatitis or encephalitis		

Levin and Hahn (1982) who treated various fulminant viral illnesses with moderate doses of interferon and showed *in vitro* recovery of the interferon system in parallel with clinical improvement in most patients. In non-fulminant viral illness, interferon was detectable in the serum of 85 per cent of patients, and in 70 per cent, the peripheral mononuclear cells were refractory to infection with vesicular stomatitis virus. Furthermore, 2′5′-oligoisoadenylate (2-5A) synthetase levels in peripheral white cells were increased above the normal range. In severely ill patients with fulminant hepatitis (caused by various different viruses including hepatitis-A, -B and herpes viruses), no interferon could be found in the serum and the peripheral mononuclear cells were sensitive to virus infection and were not able to produce interferon *in vitro* in response to virus or mitogen.

These observations are exciting and important but there is an important caveat in their interpretation. Several groups of workers have been unable to detect circulating interferon activity in the serum of patients with acute non-fulminant hepatitis, even during the pre-icteric viraemic phase. Although there has been considerable improvement in interferon assay sensitivity since these studies were made, it is important to confirm that the antiviral activity found in the non-fulminant hepatitides was interferon, now best assayed by immunoreactive methods. The therapeutic study was uncontrolled, so improvement could arguably be attributed to factors other than giving exogenous interferon. Levin argued that the deficiency in interferon was probably acquired because the patients had previously been healthy and presumably responded normally to trivial viral illness. However, we do not know whether the fulminant nature of these infections was a cause or an effect of the interferon deficiencies described.

Because the comprehensive study of the interferon system described is laborious, we do not yet have information from a detailed longitudinal study of every parameter during any natural or experimental virus infection. Such a study should clarify the temporal relationships between endogenous interferon and recovery. One early observation which suggested that interferon might be important in recovery from acute influenza was that while interferon was found in extracts of lung tissue from mice experimentally inoculated with virus and is usually found in serum and nasal washings in influenza, it could not be found in the extracts from lung tissue taken at autopsy from patients with fulminant influenzal pneumonia. even though high titres of virus were present. This study was not controlled but it has since proved possible to do more detailed longitudinal studies on the vesicular fluid of herpes zoster comparing the findings between normal individuals and cancer patients, in whom the disease is more severe (Stevens and Merigan 1972). Normally vesicle fluid from herpes zoster or herpes simplex lesions (Spruance, Green, Chiu, Yeh, Wenerstrom, and Overall 1982) contain surprisingly high titres of interferon, certainly sufficient to inhibit the replication of these viruses *in vitro,* from the earliest time that the vesicular fluid may be sampled. Individual vesicles usually crust, scab, and heal very quickly, although in any episode fresh vesicles may continue to appear in crops for a number of days. In immunosuppressed patients with a variety of tumours, particularly Hodgkin's disease, the vesicles may coalesce and remain four days without signs of rapid resolution; dissemination of the virus may also occur. These vesicles often contain undetectable or low levels of interferon. Failure to produce local interferon titres over 500 i.u. ml^{-1} correlated with general dissemination of the virus (Stevens and Merigan 1972). When interferon did appear in the vesicles (and this coincided with the appearance of varicella zoster-specific complement fixing antibody in the serum), the bases of the vesicles became more inflamed and rapid resolution followed. The failure to produce interferon in response to virus was therefore a reflection of a more general host deficiency seen in the failure to produce a prompt anamnestic response to varicella zoster antigens. This raises the question as to whether immune interferon induced soon after exposure to a reinfecting organism could be important in rapid eradication of infection. (Failure to induce immune interferon could allow the reinfection seen with respiratory syncytial virus, discussed below.) The observations with varicella were the basis for the subsequent successful trials of exogenous interferon in herpes zoster in cancer patients.

The importance of interferon in the recovery from herpes virus infection suggested by these observations in humans is confirmed by the data of Gresser, Tovey, Bandu, Maury, and Brouty-Boyé (1976) who showed that virus replication and mortality from herpes simplex virus in mice was significantly enhanced by a sheep anti-mouse interferon globulin. Paradoxically, the pathogenicity of influenza virus infection and virus titres in the same strain of mice were not enhanced by giving anti-interferon globulin parenterally or intranasally. This does not, of course, exclude an important role for interferon in recovery from influenza in man. Both natural and experimental influenza induce circulating interferon in most subjects, more so than rhinoviruses which are restricted to the upper respiractory tract, and which regularly induce local interferon only detectable in nasal washings and aspirates.

6.3.3. Development and acquisition of ability to produce interferon

There has been a suggestion that the ability to make an interferon response is not present at the time of blastocyst formation but matures slowly. There is some controversy, however, about whether cord blood cells of healthy newborns have a reduced capacity to make type I or type II interferon responses compared with older children and adults, but there is no doubt that children with congenital cytomegalovirus infection have such a deficiency. Isaacs, Clarke, Tyrrell, Webster, and Valman (1981) showed that a small proportion of wheezy infants with recurrent upper respiratory infections consistently appeared unable to make a type I interferon response either *in vivo* during the acute virus infection or *in vitro* when the peripheral blood mononuclear cells were stimulated with paramyxovirus. This did not appear to have any implications in the pathogenesis of these episodes of wheezy bronchitis (which were as protracted in the rest of the children). Virelizier, Lipinski, Tursz, and Griscell (1979) also identified a group of rare children with chronic active Epstein-Barr virus infection and deficient natural killer cell activity, who seem unable to make gamma interferon.

6.3.4. Role of host or infecting virus in the interferon response

It appears that failure to produce interferon is primarily host-determined. Relative deficiencies in either type I or type II interferon production by mononuclear cells *in vitro* have been shown in stressed mice and mice given cortisol or serotonin and in patients with marasmus, uraemia, and various tumours such as acute or chronic leukaemias and Hodgkin's disease, and in immunosuppressed patients and renal transplant recipients. Relative deficiencies of immune interferon production have been observed in some patients with leprosy (Nogueira, Kaplan, Lery, Sarno, Kushner, Granelli-Piperno, Colomer-Gould, Levis, and Steinman 1983), tuberculosis (Onwubalili and Scott 1983, and unpublished observations) and the acquired immunodeficiency syndrome (Murray, Ruben, Masur, and Roberts 1984). An increased susceptibility to, and severity and chronicity of several trivial infections in normals have been shown in many of these conditions. This does not, however, establish a cause or effect relationship with observed deficiencies in interferon production.

On the other hand, some viruses appear inefficient in stimulating a circulating interferon response *in vivo*. Acute hepatitis A or B, and glandular fever are examples although, as mentioned above, this is controversial. The enzyme 2′5′ A synthetase has been shown to be activated in acute hepatitis. It seems likely that interferon is produced in the liver, perhaps by Kupffer cells and recruited lymphocytes, and then inactivated locally.

One common cause of protracted upper and lower respiratory disease in children is respiractory syncytial virus (RSV). In several early studies, interferon was found in only 3 per cent of acute sera from patients with this infection (compared with over 50 per cent of sera from patients with influenza). Further investigations (Hall, Douglas, Simons, and Gieman 1978; Hall, Douglas, and Simons 1981; McIntosh 1978; Chonmaitree, Roberts, Douglas, Hall, and Simons 1981) confirmed that RSV induces inconsistent and low titres of interferon in peripheral mononuclear cells and in the nasal secretions of children with natural RSV infection and of adults with natural or experimental infection. Previous infection, which would have been likely in the adult volunteers, did not influence interferon production or outcome in the experimental infections and it may be concluded that interferon production is not involved in recovery from the disease. In adults RSV causes mild colds, but in infants bronchiolitis and pneumonia are common, in about 2 per cent of cases severe enough to warrant hospital admission. The disease is also unusual in that antibody induced by killed virus vaccine enhances pathogenicity, and recurrent episodes of infection with identical virus are the rule rather than the exception. The role of a lack of interferon production as a cause or effect of this unusual response may be clarified when studies are done with exogenous interferons in volunteers and patients.

6.3.5. Role of interferon in recovery from non-viral infections

Interferon production *in vivo* is not restricted to acute virus infection. A variety of Gram-positive and Gram-negative bacteria, chlamydiae, mycoplasmas, rickettsiae, and coxiella have been shown to induce interferon *in vitro* and in animals. Bacterial endotoxin induces circulating immune interferon in mice, rabbits, and man. Interferon was demonstrated in the CSF from patients with bacterial as well as aseptic meningitis. Haahr (1968) and Howie, Pollard, Kleyn, Lawrence, Peskuric, Pauker, and Baron (1982) found interferon in the ear secretions of children with acute otitis media. *Haemophilus influenzae* was the bacterial species most commonly associated with interferon in these clinical situations.

Could interferon therefore be involved in recovery from bacterial as well as viral infections? Macrophage and neutrophil phagocytosis of bacteria is enhanced and interferon is able to inhibit the invasiveness of enteric *Salmonella* spp. in tissue culture (Degré, Belsnes, Rollag, Beck, and Sonnenfeld 1983; Buckholm and Degré 1983). It remains to be seen whether these observations can be extended to animal models and perhaps even to man.

Circulating interferon has also been found in protozoan infestations; peak Type I and immune interferons occurred just before the maximum parasitaemia of murine African trypanosomiasis, and was produced with each cyclical parasitaemia. Injection of dead parasites intravenously did not induce interferon and although exogenous interferon did reduce first wave parasitaemia it did not influence subsequent mortality (Bancroft, Sutton, Morris, and Askonas 1983). Possible interference between *Plasmodium* spp. and arboviruses has long been postulated and various interferon inducers including poly I:C and bacterial endotoxin inhibit *Plasmodium berghei* in mice (Jahiel, Vilček, Nussenzweig, and Vanderberg 1968; MacGregor, Sheagren, and Wolff 1969). Interferon antibody enhances parasitaemia in the same model (Sauvaget and Fauconnier 1978) and Schultz, Huang, and Gordon (1968) were even able to show that incubating parasitized red blood cells with interferon for 6 h reduced their pathogenicity for mice! Interferon activity is detectable in the serum of patients with malarial fever (Rhodes-Feuillette, Druilhe, Canivet, Gentilini, and Périés 1981). Similarly, interferon is induced by and exogenous interferon is able to inhibit infection of tissue culture with another protozoan organism, *Toxoplasma gondii*.

6.3.6. Interferons and autoimmune disease

Interferons are found in the serum of patients with autoimmune (collagen-vascular) diseases such as systemic lupus erythematosus and Behçet's syndrome. More recently, it has been found in the serum of patients with or at risk from acquired immunodeficiency dyndrome. The source of this interferon, which appears to be acid-labile type I, neutralized by anti-leucocyte interferon and recognized by monoclonal antibodies to α-interferon, is not at all clear, nor why it is present, nor its role in the pathogenesis of disease. In the following section, the effects of exogenous interferon in man will be discussed and it should become clear that many of the nonspecific features of these autoimmune diseases can be mimicked by injecting interferons.

In summary, interferons are stimulated by a very wide range of organisms and may be important in the recovery from infections other than those caused by viruses. There are a number of diseases where interferon production is deficient and some viruses are inefficient interferon inducers. Circulating interferon is present in many diseases of obscure aetiology but not usually in normals except when they have an infection.

6.4. INTERFERON IN THE PATHOGENESIS OF DISEASE

6.4.1. Effects of exogenous interferons in animals

The large-scale manufacture of mouse and human interferons has been developed to a high degree, so we know more about the effects of these interferons in mice and men than in any other species. There is a problem with cross-species specificity, not only in their antiviral but also in their associated effects, listed in Table 6.2. Thus, although it is possible to protect

Table 6.2. *Some non-specific effects of exogenous interferons*

In suckling mice and rats
Runting, fulminant hepatic necrosis
Autoimmune glomerulonephritis
In healthy mice
Lymphopenia, granulocytosis
In humans
Inflammation
Fever and influenza-like symptoms (headache, rigors, myalgia, nausea, vomiting, diarrhoea)
Growth inhibition, weight loss, hair loss
Acute hepatitis
Acute alterations in circulating white cell components (lymphopenia and granulocytosis)
Bone marrow suppression (chronic adminstration)
Alterations in corticosteroid and lipid metabolism
Glomerulonephritis, renal transplant rejection
Metabolic changes
Central nervous system derangement
Cardiac dysrhythmias

rabbits or rhesus monkeys against vaccinia and herpes viruses using human α-interferons, it is not possible to demonstrate certain pathological changes (notably fever) that occur consistently in man. Furthermore, although rabbits will be protected against infection, they will tend to make neutralizing antibodies to human interferon. We may conclude from these observations that the cell-surface receptors for interferons in these species are similar though not identical and that there are different antigenic determinants on interferons, antibodies to which neutralize the antiviral activity. Furthermore, it would suggest that the associated effects are not mediated through the identical receptor interaction to that mediating the antiviral effect. This may be too easy an explanation, however, because rabbit interferon is not pyrogenic in rabbits and all human interferons tested so far are pyrogenic in man. Thus it is not always possible to extrapolate the findings in one animal species with homologous interferon to another.

Gresser (1982) has reviewed his experiments relating to the 'toxic' effects of interferon in mice. The most disturbing observation was of fatal fulminant hepatic necrosis in mice treated daily from birth with electrophoretically pure mouse interferon for more than seven days. The reason for this is not clear. Ultrastructural changes in hepatocyte endoplasmic reticulum were observed but their relevance is again not known. If interferon was discontinued before the seventh day after birth, the mice did not develop hepatic necrosis but they did not grow properly and at about three weeks developed immune complex type glomerulonephritis, which manifested clinically several months later. The earliest ultrastructural thickening of the glomerular basement membrane was seen by day 8 after birth and immunoglobulin and complement deposition were present by the time light microscopic changes were visible. Similar effects were seen in suckling rats treated with rat interferon but not in rabbits or rhesus monkeys treated with human leucocyte interferon.

The renal changes in lymphocytic choriomeningitis infection in mice are indistinguishable from those caused by interferon. They may in part be ascribed to endogenous interferon induction by the virus because they are proportional to the extent and persistence of interferonaemia, which varied between strains, and could be modified by anti-interferon serum.

Of interest in the possible pathogenesis of autoimmune disease in man is the observation that regular injections of either type I or immune interferons can accelerate the development of autoimmune haemolytic anaemia in inbred mice prone to such disease.

6.4.2. Effects of human interferons in man

Injections of over 10^6 i.u. of very pure human leucocyte or fibroblast interferons by any route cause dose-related fever with influenza-like symptoms (for example, headache, myalgia and malaise; reviewed by Scott 1983*a*). Repeated injections are associated with some reduction in the febrile response if the interval between injections is less than a week, but fatigue, anorexia, malaise, chronic headaches, and orthostatic hypertension supervene. The mechanisms for these symptoms are not known but it is proposed that they are not caused by interferon itself but rather by intermediate inflammatory mediators. The evidence for this is in the timing of reactions (which occur several hours after intramuscular or intradermal and between 30 and 60 min after intravenous injection) and the fact that they can be suppressed by, for example, inhibiting prostaglandin synthesis.

Intradermal injections of interferons cause local inflammation beginning 2-4 h after injection, maximal at 8–24 h. It has now been demonstrated that interferon (fibroblast interferon at least) can stimulate inflammatory prostaglandin synthesis *in vitro*. However, when the symptoms and febrile response to interferons are suppressed by indomethacin (a potent cyclo-oxygenase inhibitor), various other metabolic changes remain unaffected. These include a hydroxycorticosteroid response and changes in circulating trace elements. It is not yet known whether interferon stimulates corticosteroid production at the adrenal cortex, at the pituitary or the hypothalamus. A related phenomenon may be the reduction in circulating high-density lipoproteins seen in volunteers treated with about 3×10^6 i.u. leucocyte interferon daily for four days. It is of interest to note that corticosteroids and stress inhibit interferon production and interfere with its antiviral action.

Interferon also stimulates IgE-mediated degranulation of sensitized mast cells. This has not yet been proven to relate directly to the phenomenon that trivial virus infections are often associated with exacerbation of asthma or wheezy bronchitis. In normals, however, injections or inhalations of interferon do not cause bronchospasm, though inhalation by small particle aerosol of sufficient monoclonal antibody-purified leucocyte interferon to cause a febrile systemic reaction ($> 10^7$ i.u.) caused increased alveolar-capillary permeability (Heneghan and Scott, personal observation).

Occasionally, repeated injections of interferon do cause allergic phenomena such as local delayed hypersensitivity or generalized urticaria or non-specific erythematous rashes but it is not yet known whether these are due to repeated injections of some allergenic impurity.

6.4.3. Interferon effects and viral disease symptoms

All the 'toxic' or associated features of interferons described so far occur at a dose level which would give circulating or local tissue levels expected in virus diseases. After intramuscular injection of 3×10^6 i.u. leucocyte interferon, peak serum levels of 10–100 i.u. ml^{-1} are seen. Similar levels are found in influenza, for example, where the symptoms experienced are the same. In the vesicles of herpes simplex and herpes zoster, much higher titres may be found (sometimes over 10^4 i.u. ml^{-1}). Intradermal injection of such amounts causes local inflammation.

More striking toxic effects have been seen with very high doses of interferons given by continuous intravenous infusion for the treatment of tumours over a period of several days or weeks. It is doubtful that these levels are ever achieved in common virus infections although very high titres (mean 5×10^3 i.u. ml^{-1}) have been found in Argentinian haemorrhagic fever (Levis, Saavedra, Ceccoli, Falcoff, Feuillade, Enria, Maiztegui, and Falcoff 1983). Perhaps this is characteristic of virus illness with a prolonged viraemic phase, although very high titres were unusual in the acute phase of Lassa fever (Scott, Robinson, and McCormick, unpublished observations). The best way to stimulate high circulating titres of interferon in experimental animals is to inject a good inducer (such as poly rI. poly rC or Sendai virus) intravenously. In humans, the effects of poly rI. poly rC are indistinguishable from interferon but arguably just as non-specific.

In high doses, α-interferons cause severe malaise with central nervous system disturbance (confusion, stupour or coma) and excessive slow-wave abnormalities on electroencephalogram (EEG). There may be marked metabolic disturbances with hyperkalaemia and occasional hypocalcaemia. Hyperglycaemia and minor rises in urea and creatinine have also been described. Occasionally, disturbances in cardiac rhythm have been seen, at least one patient has died in coma and there have been small numbers of sudden unexpected deaths, particularly in patients with previous heart disease. It is of considerable interest whether these rare events seen with exogenous interferons reflect changes sometimes and inexplicably seen with natural infections. For example, in measles and rubella and even after live measles vaccination in children, unexplained confusion with slow-wave EEG abnormalities are often seen.

Moderate or high doses of interferon also cause rises in aspartate transaminase, thought (in the absence of electrocardiographic changes) to be of hepatic origin. The treatment of chronic active hepatitis with interferon may also cause a rapid short-lived rise in this enzyme. With long-term daily administration of moderate doses of interferon, there are also rises in other liver-associated enzymes, γ-glutamyl transpeptidase and alkaline phosphatase. The dose of interferon required for severe changes in liver function in cancer patients was around 10^8 i.u. per day, and it was estimated that the dose of interferon equivalent to that needed to cause hepatic necrosis in newborn mice was approximately 10^7 i.u. kg^{-1} day^{-1}. Doses considerably lower than this caused failure of weight gain in children with congenital cytomegalovirus infection and the repeated doses over 3×10^6 i.u. caused weight loss in adults, perhaps in part by suppressing appetite. However, in adults, there is other evidence of a direct inhibition of cell growth, the most obvious being hair loss.

Doses of interferon even below those needed to cause a febrile reaction (i.e. $< 10^6$ i.u.) always cause lymphopenia and transient neutrophil leucocytosis in volunteers and patients. Lymphopenia is a very common accompaniment to acute virus infections in humans, and Schattner, Meshorer, and Wallach (1983) concluded that the lymphopenia seen when vesicular stomatitis virus was given to mice intraperitoneally was due to endogenous interferon, although the transient granulocytosis which was also seen was not inhibited by anti-interferon serum. Higher doses ($> 10^6$ i.u. day^{-1}) and prolonged treatment cause reversible suppression of all components of the bone marrow. Granulocytopenia and thrombocytopenia are common reasons for discontinuing or modifying the dosage regimens in clinical trials, particularly in the treatment of conditions where the marrow may already be affected (as in chronic cytomegalovirus infection) or when other agents (such as cotrimoxazole or immunosuppressives) are being given.

Some toxic effects in man appear to be idiosyncratic. A sporadic case of actue interstitial nephritis was seen after doses of 10^8 i.u. IFN-αA were given three times per week for three weeks (Averbuch, Austin, Sherwin, Antonovych, Bunn, and Longo 1984). More worryingly, IFN-αA in doses of 3.6×10^6 i.u. three times per week was associated with renal transplant rejection in all of eight patients compared with only one of eight on placebo. The glomeruli were spared in most of these patients but there was interstitial inflammation (Kramer, ten Kate, Bijnen, Jeekel, and Weimar 1984). Though not exactly comparable because of difficulty in standardizing assays, this dose is only a little over that of natural leucocyte interferon shown by Hirsch, Schooley, Cosimi, Russell, Delmonico, Tolkoff-Rubin, Herrin, Cantell, Farrell, Rota, and Rubin (1983), to inhibit cytomegalovirus infections in renal transplant recipients.

The evidence that endogenous interferon mediates any of the pathogenic changes in virus infections of man is only circumstantial. For example, in elegant studies of interference between live attenuated measles virus and subsequent vaccination, Petralli, Merigan, and Wilbur (1965) showed that the optimum protection against vaccinia virus occurred at a time when children were febrile. Interferon is often found in the sera or nasal washings of patients with febrile viral illnesses such as influenza but in only a small proportion (perhaps 20 per cent) of those with febrile illness suspected to be of viral origin, but in which an agent is not detected. In contrast, respiratory syncytial virus colds in which interferon is rarely found are usually free of the severe general systemic reactions characteristic of influenza. Interferon is found locally in high titre in vesicles of herpes simplex and varicella zoster and may contribute to local inflammation: it is also found in brain biopsy extracts of patients with herpes simplex encephalitis, but it is not known if there are similar high titres in the brain in measles and rubella when disturbances of behaviour and consciousness and abnormal EEGs are observed. Local interferon may be produced in very high titre in organs where virus is replicating, so local inflammation could be attributed, at least in part, to this interferon. Thus the rhinitis and coryza of rhinovirus colds, the parotitis and pancreatitis of mumps and the hepatitis of hepatotropic virus infections may in part be caused by locally induced interferon.

If interferon is in fact involved, it seems likely that inflammation is multifactorial and mediated by chemical substances (interleukins, prostaglandins, leukotrienes, histamine, 5-hydroxytryptamine, etc.) which may be produced in response to many lymphokines and cytokines. The focus on interferons has occurred simply because they are the first of this large group of substances to have been well-defined, manufactured in large amounts to a high degree of purity and actually studied in normal volunteers as well as patients, in whom any interpretation of pathogenesis may be suspect.

6.5. THE USE OF INTERFERON TO PREVENT AND TREAT VIRUS INFECTION

There has been almost no work on the potential use of interferon for the treatment of infections other than those of viral aetiology. Some diseases of uncertain aetiology (for example, Crohn's disease, Behçet's syndrome, Jakob-Creutzfeldt disease, and multiple sclerosis) have been treated either on the premise that they are caused by viruses or that they are immunologically mediated. Most of these studies have been uncontrolled and for the most part, no benefits could be demonstrated. That circulating interferon is found in autoimmune disease in man, and accelerates autoimmune disease in inbred mice, suggests that it may be unhelpful or unwise to give more exogenous interferon in these cases, but only controlled trials will prove this. Trials of interferon in certain bacterial or parasitic infections may certainly be worthwhile. Many deficiencies, for example in cell-mediated immunity or phagocytosis in tuberculosis and leprosy could perhaps be corrected by modest doses of exogenous interferon. Caution should be taken, however, as an exuberant immunological response to extensive infection may be detrimental, as was probably the case in the rejected renal transplants in interferon recipients. It is clear that even appropriate bactericidal chemotherapy is sometimes ineffective in the absence of intact host defence, and immunological adjuvants to standard antibacterial therapy are being considered. One of the most exciting secondary observations to come out of any trial of interferon has been of a possible reduction in infections other than those caused by viruses in renal transplant recipients (Hirsch *et al.* 1983). This could come about by a direct effect of interferon inhibiting replication, by enhancing macrophage activity and other immune functions or by preventing the expression of immunosuppressive viruses, particularly cytomegalovirus.

In contrast, many trials have been performed with exogenous type I interferons and interferon inducers in virus diseases (Table 6.3, reviewed by Scott 1983*b*). Because interferon is normally produced early in acute virus infection, at a time when symptoms have begun, it is generally felt (but not proven) that further exogenous interferon cannot affect the normal rapid resolution of most such infections. This is particularly so if the symptoms are, as discussed above, in part caused by endogenous interferon. To be effective therefore, interferon has either to be given before infection, or in acute infections which are protracted because endogenous interferon is not made (Table 6.1), or in chronic infections where the endogenous interferon, if it is made at all, is incapable of eradicating infection. The effects of interferon are enhanced by combination with chemical antiviral agents for herpes simplex eye disease and possibly for chronic active hepatitis; this approach to effective antiviral therapy will be important in future clinical studies.

6.5.1. Sources of interferon for clinical use

Most studies have been done with partially purified leucocyte interferon induced by paramyxovirus in pooled buffy-coat leucocytes from blood for transfusion. Many of the observations made with this material have now been confirmed with this material purified to homogeneity and with individual leucocyte interferon species derived by recombinant DNA synthesis in *E. coli,* and with fibroblast and lymphoblastoid interferons from tissue culture. Thus, many of the early suspicions that there could be undefined ingredients in interferon preparations which were essential for antiviral

Table 6.3. *Use of exogenous interferons in man*

Viral diseases modified or prevented by prophylactic interferon
In volunteers
Vaccinia skin lesions
Rhinovirus colds
Coronavirus colds
Influenza
Rubella vaccine
In patients
Reactivation herpes simplex labialis (trigeminal surgery)
Reactivation cytomegalovirus (renal transplant recipients)
Acute viral diseases which respond to interferon
Herpes simplex dendritic keratitis
Varicella zoster (shingles or chicken-pox in immunocompromised patients)
Chronic viral diseases which respond to interferon
Cytomegalovirus (transient response)
Hepatitis B virus-associated chronic active hepatitis
Viral papillomas
Responses claimed in poorly controlled studies
Acute adenovirus conjunctivitis
Wild upper respiratory infection
Vaccinia keratitis
Herpes simplex (primary and secondary infections)

effects *in vivo* have been allayed. Paradoxically, it has also become clear that the toxic effects discussed above and listed in Table 6.2 are properties of interferons themselves and these toxic effects are a serious limitation to the exuberant prescription of very large doses of interferon for any condition. At least, there is now no longer any restriction on the supply of interferon for clinical trials, although more of the recent production by successful pharmaceutical companies has been directed towards cancer than virus infections. Most of the studies described below were done with very small amounts of interferon.

There is insufficient data on efficacy as yet to allow the granting of a product licence to market a particular type of interferon for any one indication in the UK or USA. Detailed toxicity studies with human interferons probably can only be done in humans, because of cross-species specificities. This is a unique situation for regulatory agencies. Nevertheless, interferon has been freely available from pharmacists in the USSR and Yugoslavia for many years: in the former as nose drops for the treatment of upper respiratory infections and in the latter as a cream for the treatment of genital warts. Both preparations are of low titre and it is doubtful that the former preparation has any antiviral effect at all, though a strong placebo effect could not be denied. In 1983, partially purified fibroblast interferon was given a product licence in Germany for the treatment of herpes zoster in immunocompromised patients. However, no properly controlled clinical trial data have been published in support of this apparent claim for efficacy. The results of trials with leucocyte interferon cannot be assumed to apply to fibroblast interferon because there are important differences in pharmacokinetics between the types: to achieve equivalent serum antiviral levels after intramuscular dosing, more fibroblast interferon is needed, probably because it is released more slowly from intramuscular injection sites. The two established acute illnesses where interferon has been proven to have an effect are herpes simplex keratitis and herpes zoster in immunocompromised patients. These conditions are relatively uncommon, collection of data is slow, so it may be some time before interferon is freely available for treatment. From the regulatory authority point of view, the situation is more complex because combination therapy with nucleoside analogue antiviral agents will probably be more effective than interferon alone, and combinations have enhanced toxic effects.

6.6. DISEASES MODIFIED BY EXOGENOUS INTERFERONS

6.6.1. Volunteer trials

6.6.1.1. Vaccinia virus

Before Isaacs discovered interferon, he had shown that injections of inactivated influenza virus could inhibit vaccinia lesions in rabbit skin (Depoux and Isaacs 1954). It was natural that when the antiviral substance was defined, similar experiments would be done. Although chick embryo interferon was not very active in this model, much better protection was seen using crude low-titre rabbit and monkey kidney cell interferons. It was then shown that monkey kidney interferon given intradermally could protect skin sites against vaccinia when vaccination was performed 24 h later, and subsequently this experiment was repeated using human fibroblast interferon. Animal models of vaccinia have been used to compare various preparations of human interferon (Weimar, Stitz, Billiau, and Schellekens 1980). Because human interferons work well against vaccinia in rhesus monkeys, but not so well *in vitro,* Schellekens, Weimar, Cantell, and Stitz (1979) suggested that the effects against vaccinia may not be entirely due to interferon's antiviral action but rather through some host-mediated effect, presumably through the immune system. It is also of interest how much easier it is to inhibit vaccinia than herpes simplex lesions in animals with low dose interferon. Petralli *et al.* (1965) showed that measles vaccination could protect against vaccinia inoculated on days 9–15, at a time when circulating interferon was most commonly found. Vaccinia virus itself induces a good local interferon response (Wheelock 1964) and has been used in the therapy of warts, although this method of treatment never became established.

Wheelock proposed that local interferon generated by variola in the skin prevented the dissemination and development of smallpox. From what we now know about the pathogenesis of herpes zoster in immunocompromised patients, this may be the case.

6.6.1.2. Common colds

Early studies against a variety of agents which caused colds in volunteers showed no effect, probably because the titre of interferon given intranasally was too low, but also because too few volunteers were tested to show a modest response. It took 15 years from the discovery of interferon to develop the manufacturing process to enable sufficient leucocyte interferon to be made over about one year to do limited but well controlled trials against influenza and rhinovirus colds at the MRC Common Cold Unit (Merigan, Reed, Hall, and Tyrrell 1973). Repeated doses of interferon over 24 h before influenza challenge delayed the symptoms slightly. By continuing interferon over a total of four days and challenging with rhinovirus on day 2, a definite protective effect against severe colds could be shown. Apart from small trials with interferon inducers and low-titre fibroblast interferon, no further work was done at the Common Cold Unit until 1980, when monoclonal antibody-purified leucocyte interferon and recombinant DNA IFN-α_2 were both shown to be highly effective in the same rhinovirus model. In these early trials, excessive interferon was given to be sure of an effect and the minimal dose requirements to prevent colds consistently have now been established (Phillpotts, Scott, Higgins, Wallace, Tyrrell, and Gauci 1983). A single intranasal spray of high concentration (dose 3×10^6 i.u.) interferon given daily from one day before virus challenge will protect, providing there is not a long interval between interferon and virus challenge. When the latter was delayed until late in the evening on the second day, no protection at all could be demonstrated. This is a peculiar result because in theory the antiviral state induced by interferon, if effective, should persist for many hours. The nasal epithelial cells susceptible to rhinovirus (and at present, we do not know exactly which cells these are) probably have to be in a particularly resistant state to inhibit very early events in virus invasion in order to prevent successful infection and symptoms.

The conclusion from these dose-ranging studies was that interferon needed to be given three times a day (at about 3×10^6 i.u. per dose), in order to protect volunteers consistently against rhinovirus challenge at any time of day. This dose of IFN-αA was also found to be effective against coronavirus colds (Higgins, Philpotts, Scott, Wallace, Bernhardt, and Tyrrell 1983) and a slightly different dose schedule (5×10^6 i.u. twice a day) was shown by Dolin, Betts, Treanor, Erb, Roth, and Reichman (1983) to partially protect volunteers against influenza A challenge. However, it has transpired that the dose of any interferon cannot be reduced further and still protect against wild colds.

When interferon was still in short supply, a series of experiments was performed to determine why so much intranasal interferon appeared to be necessary to prevent colds (Harmon, Greenberg, and Couch 1976; Harman, Greenberg, Johnson, and Couch 1977). First there are pharmacokinetic problems. Intranasally applied substances are cleared very rapidly from the nose, although in recent experiments it has been shown that low titres of interferon are still recoverable by nasal washing 24 h after a single dose. In addition, nasal secretions tend to inactivate human interferons, particularly natural fibroblast interferon *in vitro,* although this is less important in the case of leucocyte interferon. Secondly, interferon (at 1000 i.u. ml^{-1}) needed to be incubated with explanted nasal epithelial cells for 4 h in order to protect against vesicular stomatitis virus. (Rhinoviruses do not grow in adult nasal epithelial organ culture.) Alternative methods of applying interferon were tested by Greenberg, Harmon, Couch, Johnson, Wilson, Dacso, Bloom, and Quarles (1982). Continuous intranasal sprays or application by cotton wool pledget both seemed effective against rhinoviruses in the doses tested. However, there is no simpler method of giving a nasal preparation than by infrequent single sprays of high titre material, and with the advent of almost unlimited supplies this method has now been adopted for large-scale trials of prophylactic interferon against colds.

Unfortunately, the dose of interferon needed to protect against wild colds has been found to cause local inflammation, with nasal discomfort and nose-bleeding occurring after one or two weeks' regular application. Superficial nasal mucosal biopsies after 28 days' intranasal IFN-α_2 showed marked subepithelial mononuclear cell infiltration with some ulceration (Hayden, Mills, and Johns 1983). The symptoms appear to occur irrespective of the preparation and purity of the interferon and of the preservatives used in leucocyte interferons tested so far. Further trials with IFN-α subtypes, IFN-β, and IFN-γ are needed to determine whether this property is consistent and correlates with antiviral activity. If this proves to be the case, then it is unlikely that interferon alone will be used for prophylaxis against colds or even influenza on a wide scale, even though it is highly effective.

An alternative approach would be to give interferon for a short period to family contacts of cold cases in order to prevent the disease. In a small trial IFN-αA has already been shown to have some beneficial effect (Herzog, Just, Berger, Havas, and Fernex 1983), and larger more definitive trials with confirmatory virology are under way.

There have been several studies from the USSR, Bulgaria, Yugoslavia, and Japan in which beneficial responses to therapeutic or prophylactic low dose intranasal interferon were claimed for colds and influenza. With our present knowledge of the amount of interferon needed to prevent natural and experimental colds, suspicions about the results of these trials have hardened.

6.6.2. Patient trials

6.6.2.1. Prevention of reactivation herpes simplex labialis

Herpes simplex virus is latent in trigeminal ganglia in patients who tend to reactivation herpes simplex labialis (cold sores). Mobilization of the trigeminal ganglia or root for the treatment of tic doloureux often gives rise to cold sores and this may be aggravated by dexamethasone used to reduce local oedema and inflammation at the operation site. In an elegant study, Pazin, Armstrong, Lam, Torr, Janetta, and Ho (1979) showed that interferon given intramuscularly from the day before operation could reduce the frequency of cold sores and of oropharyngeal herpes simplex virus secretion by about half, at the expense of not inconsiderable toxicity. This treatment, however, did not have any effect on subsequent attacks of cold sores.

Paradoxically, one interesting aspect of the toxic febrile reaction to interferon seen in volunteers given a single dose is reactivation herpes simplex beginning at 48 h! A single dose is therefore insufficient (probably not persistent enough) to prevent virus replication even though it may stimulate reactivation of latent virus, perhaps by inflammatory prostaglandin synthesis.

6.6.2.2. Prevention of viral syndromes in renal transplant recipients

An early trial with intramuscular fibroblast interferon given twice a week for three months after renal transplantation failed to show any benefit in terms of suppression of cytomegalovirus or of other common acute virus infections (Weimar, Schellekens, Lameijer, Masurel, Edy, Billiau, and De Somer 1978). The same dose of leucocyte interferon did, however, delay the onset of cytomegalovirus (CMV) excretion (Cheeseman, Rubin, Stewart, Tolkoff-Rubin, Cosimi, Cantell, Gilbert, Winkle, Herrin, Black, Russell, and Hirsch 1979; Cheeseman, Henle, Rubin, Tolkoff-Rubin, Cosimi, Cantell, Winkle, Herrin, Black, Russell and Hirsch 1980) and suppress Epstein-Barr virus shedding. When leucocyte interferon was given more frequently and for longer (Hirsch *et al.* 1983), the previous observations were confirmed and, in addition, syndromes due to CMV reactivation such as pneumonia, were significantly suppressed. The toxicities were as previously described, thrombocytopenia being the most common reason for stopping treatment, but renal transplant recipients tend to thrombocytopenia for a variety of other reasons anyway.

In patients with congenital CMV infection, transient reductions of cytomegaloviruria have been observed with interferon as well as with nucleoside analogues. No consistent clinical benefit of any treatment has been convincingly documented.

6.6.2.3. Treatment of acute cytomegalovirus syndromes

Whereas it looks as though leucocyte interferon probably prevents or delays

CMV reactivation in renal transplant recipients, neither interferon nor nucleoside analogues nor combinations of both have any therapeutic value in established cytomegalovirus infection in bone marrow transplant recipients. Meyers and coworkers (Meyers, McGuffin, Nieman, Singer, and Thomas 1980; Meyers, McGuffin, Bryson, Cantell, and Thomas 1982) first examined leucocyte interferon alone and then interferon combined with adenine arabinoside in open trials and most recently IFN-αA alone (Meyers, Day, Lum, and Sullivan 1983). Most of these patients died despite a hundred fold fall in the titre of CMV between diagnostic lung biopsy and post-mortem lung tissue extracts. The combination therapy caused significant neutropenia, thrombocytopenia, and neurotoxicity. Similar toxicities were seen with a combination of leucocyte interferon and acyclovir (Wade, Hintz, McGuffin, Springmeyer, Connor, and Meyers 1983). Although three of thirteen patients survived the disease and the treatment, the effect on cytomegalovirus titres in lung extracts was less impressive than with leucocyte interferon alone. Although uncontrolled, these trials do indicate that the prognosis for CMV infections in bone marrow transplant recipients is very poor. The unwanted effects of interferon and nucleoside analogues are additive and are a particular disadvantage in this group of patients in whom bone marrow function is already compromised.

6.6.2.4. Treatment of herpes simplex keratitis

Early trials in experimental herpes (HSV) keratitis in animal models and against acute disease in man were unsuccessful until titres of interferon over 10^5 i.u. ml^{-1} were tested. The 50 per cent inhibitory concentration of human fibroblast interferon against HSV in African green monkey eyes was 1.9×10^5 i.u. ml^{-1} (interferon being given 15 h before and with the virus) and at this concentration leucocyte interferon was slightly more effective (Neumann-Haefelin, Sundmacher, Skoda, and Cantell 1977). Although 6.23×10^4 i.u. ml^{-1} human leucocyte interferon failed to accelerate healing in acute dendritic keratitis, a higher concentration (3×10^6 i.u. ml^{-1}) did accelerate healing after thermocautery (Sundmacher, Neumann-Haefelin, and Cantell 1976), and after minimal debridement (Jones, Coster, Falcon, and Cantell 1976). The effects of concentrations of 10^6 i.u. ml^{-1} leucocyte and fibroblast interferons were not different (Sundmacher, Cantell, Skoda, Hallermann, and Neumann-Haefelin 1978).

Several groups have now shown that healing rates of herpetic ulcers can be enhanced by giving combinations of interferon with trifluorothymidine or with acyclovir compared with either alone (see review by Sundmacher 1982). The results of a dose-ranging study suggests that there is no advantage in giving leucocyte interferon eye drops at a concentration higher than 30×10^6 i.u. ml^{-1} (Sundmacher, Cantell, Mattes 1984).

Thus interferon eye drops can accelerate the healing of acute herpes simplex dendritic ulcers. This may be because interferon is not normally made in the cornea in response to such infection. Studies in animals and man suggest that it is quite sufficent to give interferon eye drops once a day, and although high concentrations might be expected to cause local inflammation if treatment is prolonged, this has not yet been described. Eye drops of nucleoside analogues usually need to be given more frequently than once a day so it may be impractical to combine the two in one medication.

Herpes keratoconjunctivitis like herpes labialis tends to recur, but trials of regular interferon eye drops have so far failed to prevent recurrences. Deeper keratitis and stromal infection of the cornea often arise with chronic or recurrent HSV infection because of exuberant hypersensitivity reaction in the stroma. This is best treated by a combination of corticosteroids and antiviral agents and interferon may well have a role. Unfortunately, trials in these patients are difficult because the cases are uncommon, the disease is not homogenous and the results of therapy are very difficult to assess. It must be accepted that in certain circumstances empirical therapy will be adopted if it does not appear to cause harm.

6.6.2.5. Varicella zoster: chicken-pox and shingles in immunocompromised patients

In patients with tumours, especially Hodgkin's disease, varicella zoster infections are more severe and prolonged. As discussed above, delay in natural resolution of shingles vesicles is associated with a defect in local interferon and systemic antibody production. Merigan, Rand, Pollard, Abdallah, and Jordan (1978) showed a convincing reduction of systemic spread of the virus, local spread in the primary dermatome, and complications of infection with interferon initially at 2.55×10^5 i.u. kg^{-1} day^{-1} given 12-hourly and reducing over a period of 5 days. The benefit was less clear when treatment was discontinued after four doses (Merigan, Gallagher, Pollard, and Arvin 1981). Arvin, Kushner, Feldman, Baehner, Hammond, and Merigan (1982) found that slightly higher doses (3.5×10^5 i.u. kg^{-1} day^{-1} for three days) were needed to inhibit chicken-pox in immunosuppressed children in the same way. However, one child died of varicella pneumonia after treatment had been discontinued. It is likely that higher doses will not be well tolerated though it should be possible to continue treatment for longer than in the trial (3 days) in an attempt to prevent the occasional persistent infection or reactivation. Because adenine arabinoside and acyclovir are both active against varicella in immunosuppressed children, it is no longer ethical to do placebo-controlled trials. Combinations may be more effective if not also more toxic.

6.6.2.6. Chronic hepatitis B virus (HBV)-associated chronic active hepatitis

Following acute hepatitis B, about 10 per cent of patients go on to chronic carriage of HBV surface antigen (HBsAg) but many fewer have chronic hepatitis. This disease is classified histologically (as 'active' or 'persistent' hepatitis) and by reference to circulating markers of HBV infection. HBV-associated DNA polymerase (DNAP) and HBeAg are markers of infectivity and continuing HBV replication. Many uncontrolled trials have now been done with interferons and nucleoside analogues in these patients. There is no doubt that interferon can inhibit HBV replication as measured by consistent falls in HBV DNAP. Unfortunately, in the majority this fall is only transient. In a small proportion of patients HBV DNAP, HBeAg, and even HBsAg (which is a surface coat protein produced in excess of requirements to make complete virus particles) may be cleared from the serum; the patients improve clinically and liver biopsy appearances improve. Courses of interferon combined with adenine arabinoside showed benefit in up to half of the patients treated, compared with a quarter or less of those treated with either alone. To get round problems of toxicity, intermittent courses of treatment were given (Scullard, Pollard, Smith, Sacks, Gregory, Robinson, and Merigan 1981*a*; Scullard, Andres, Greenberg, Smith, Sawhney, Neal, Mahal, Popper, Merigan, Robinson, and Gregory 1981*b*). Improvement in these patients must be interpreted in the light of observations that HBV DNAP falls spontaneously in about 10 per cent of patients with chronic active hepatitis per year. The natural history of the disease is not clear and future trials need to be adequately controlled to confirm that the treatment (which causes some discomfort and toxicity) is beneficial.

6.6.2.7. Treatment of viral papillomas (warts)

In Yugoslavia, crude leucocyte interferon cream of titre approximately 10^4 i.u. g^{-1} applied liberally to genital warts in men and women over several weeks caused most of them to resolve. Much higher titre cream has now been tried in refractory vaginal flat condylomata with success (Vesterinen, Meyer, Purola, and Cantell 1984). Intralesional injections of leucocyte interferon also cure body warts (verruca vulgaris) but the optimal treatment course has yet to be established. Systemic interferon does not appear to affect body warts although intramuscular fibroblast interferon does induce resolution of genital warts (Schonfeld, Nitke, Schattner, Wallach, Crespi, Hahn, Levani, Yarden, Shoham, Doerner, and Revel 1984). This difference in response between different types of warts probably reflects the fact that they are caused by different strains of papillomavirus with very different modes of transmission and pathogenesis. Body warts contain very large amounts of virus with almost no cellular infiltration whereas in genital warts, virus is scarce and there is moderate oedema with mononuclear cell infiltrate. Further studies with interferon, which could work by enhancing cell-mediated immunity as well as by inhibiting papillomavirus replication, may help to unfold the pathogenesis of these two conditions.

6.6.2.8. Acute adenovirus conjunctivitis

A comparison between fibroblast interferon eye drops and local corticosteroids showed improved healing rates with the former during an epidemic of adenovirus conjunctivitis in Israel (Romano, Revel, Gurari-Rotman, Blumenthal, and Stein 1980). It is probable that this difference was in part due to prolongation of illness by corticosteroids. Preliminary double-blind trials with the sort of doses of leucocyte interferon known to inhibit herpes simplex *in vivo* suggested no benefit in established adenovirus conjunctivitis although some protection against autoinoculation cross-infection was suggested (Sundmacher, Wigand, and Cantell 1980). Further trials are indicated during epidemics of this disease but it is impractical to do studies in sporadic cases of infection with different serotypes of adenovirus.

6.7. SUMMARY AND CONCLUSIONS

What impact has the study of the interferon system had on progress in understanding and treatment of infectious diseases? In this chapter, some evidence has been summarized which suggests that interferon is produced in many acute viral illnesses and is probably important in the rapid resolution (and hence the triviality) of the vast majority of illnesses suffered by man. We can define situations where interferon does not appear to be produced as usual and show that viral illnesses are then more severe and prolonged. Note, however, that cause and effect have not been established. Interferon production is not restricted to viral illness and broader immune functions may be just as important as the antiviral action in recovery from infections. There may also be a role for continuous or periodic low-level interferon production in lymph tissue as a 'regulator' of immune responsiveness.

On the other hand, interferons also have quite dramatic 'toxic' effects in man. Whereas they probably are not the exclusive or prime mediators of inflammation and fever in virus illness, they are strong contenders for an important role. It seems likely that interferons act by stimulating endogenous pyrogen production (perhaps interleukin I) which itself causes the release of inflammatory mediators although a direct effect of interferons has yet to be excluded. The extraordinary effects seen when given in moderate or high doses, though non-specific, are very like the effects of certain acute viral illnesses or malarial fever. Paradoxically, in the worst syndromes of fulminant hepatitis, interferon was not detectable.

From its beginnings as a prospective broad-spectrum 'penicillin' for the treatment of virus infections, interferon does appear to have found a role in certain restricted clinical situations. The most important benefits can be shown when there is a deficiency of normal interferon production or when interferon is given before infection. It is possible to protect against common colds but at the risk of causing local symptoms if treatment is continued for too long. Perhaps local interferon produced in high concentration in the mucosa early in colds contributes to the rhinitis.

It is possible to protect against cytomegalovirus syndromes in renal transplant recipients but slightly higher doses are toxic and may cause transplant rejection. Thus the therapeutic ratio for interferon in virtually every important clinical situation approaches unity. Exceptions to this are the topical use of interferon eye drops for herpes keratitis and the local treatment of warts.

There is no doubt that interferons will become available (in the West) for the treatment of diseases of man in the near future. When they are first marketed for restricted indications, it is inevitable that they will be used more freely by physicians in the treatment of other unrelated diseases. Because interferons are toxic, it is only by carefully controlled trials that the real benefits can be properly assessed. It is to be hoped that such assessments are made, not only to establish therapeutic value, but also to prevent disillusionment, mostly because of toxicity. Perhaps the most exciting application will be of 'low' dose interferon as an immune modulator in a wide range of different infections. The discoveries of interferons and their ramifications throughout the immune system have strikingly improved our understanding of all aspects of pathogenicity and host defence. The hypothetical nature of many comments in this chapter suggest that there is still much to do.

6.8. REFERENCES

Arvin, A. M., Kushner, J. H., Feldman, S., Baehner, L., Hammond, D., and Merigan, T. C. (1982). Human leukocyte interferon for the treatment of varicella in children with cancer. *N. Engl. J. Med.* **306,** 761–5.

Averbuch, S. D., Austin, H. A., Sherwin, S. A., Antonovych, T., Bunn, P. A., and Longo, D. L. (1984). Acute interstitial nephritis with the nephrotic syndrome following recombinant leukocyte A interferon therapy for mycosis fungoides. *N. Engl. J. Med.* **310,** 30–1.

Bancroft, G. J., Sutton, C. J., Morris, A. G., and Askonas, B. A. (1983). Production of interferons during experimental African trypanosomiasis. *Clin. Exp. Immunol.* **52,** 135–43.

Bocci, V. (1981). Production and role of interferon in physiological conditions. *Biol. Rev.* **56,** 49–85.

—— Muscettola, M., Paulesu, L., and Grasso, G. (1984). The physiological interferon response II. Interferon is present in lymph but not in plasma of healthy rabbits. *J. Gen. Virol.* **65,** 101–108.

Buckholm, G. and Degré, M. (1983). Effect of human leukocyte interferon on invasiveness of salmonella species in Hep-2 cell culture. *Infect. Immun.* **42,** 1998–2202.

Cheeseman, S. H., Henle, W., Rubin, R. H., Tolkoff-Rubin, N. E., Cosimi, B., Cantell, K., Winkle, S., Herrin, J. T., Black, P. H., Russell, P. S., and Hirsch, M. S. (1980). Epstein-Barr virus infection in renal transplant recipients. *Ann. Intern. Med.* **93,** 39–42.

—— Rubin, R., Stewart, J., Tolkoff-Rubin, N., Cosimi, A., Cantell, K., Gilbert, J., Winkle, S., Herrin, J., Black, P., Russell, P., and Hirsch, M. (1979). Controlled clinical trial of prophylactic human leukocyte interferon in renal transplantation. *N. Engl. J. Med.* **300,** 1345–9.

Chonmaitree, T., Roberts, N. J., Douglas, R. G., Hall, C. B., and Simons, R. L. (1981). Interferon production by human mononuclear leucocytes: differences between RSV and influenza. *Infect. Immun.* **32,** 300–3.

Degré, M., Belsnes, K., Rollag, H., Beck, S., and Sonnenfeld, G. (1983). Influence of the genotype of mice on the effect of interferon on phagocytic activity of macrophages. *Proc. Soc. Exp. Biol. Med.* **173,** 27–31.

Depoux, R. and Isaacs, A. (1954). Interference between influenza and vaccinia viruses. *Br. J. Exp. Pathol.* **35,** 415–8.

Dolin, R., Betts, R. F. Treanor, J., Erb, S., Roth, F. K., and Reichman, R. C. (1983). Intranasally administered interferon as prophylaxis against experimentally induced influenza A infection in humans. *Proceedings of the 13th International Congress of Chemotherapy, Vienna,* Vol. 6, pp. SE4.7/1–7. Verlag Ergermann.

Greenberg, S. B., Harmon, M. W., Couch, R. B., Johnson, P. E., Wilson, S. Z., Dacso, C. C., Bloom, K., and Quarles, J. (1982). Prophylactic effect of low doses of leukocyte interferon against infection with rhinovirus. *J. Infect. Dis.* **145,** 542–6.

Gresser, I. (1982). Can interferon induce disease? In *Interferon 4,* (ed. I. Gresser) pp. 95–127. Academic Press, London.

—— Tovey, M. G., Bandu, M.-T., Maury, C., and Brouty-Boyé, D. (1976). Role of interferon in the pathogenesis of virus diseases in mice as demonstrated by the use of anti-interferon serum. *J. Exp. Med.* **144,** 1305–15.

Haahr, S. (1968). The occurrence of interferon in the cerebrospinal fluid in patients with bacterial meningitis. *Acta Pathol. Microbiol. Scand.* **73,** 264–74.

Hall, C. B., Douglas, R. G., and Simons, R. L. (1981). Interferon production in adults with respiratory syncytial virus infection. *Ann. Intern. Med.* **94,** 53–5.

——, —— Simons, R. L., and Geiman, J. M. (1978). Interferon production in children with respiratory syncytial, influenza and parainfluenza virus infections. *J. Paediatr.* **93,** 28–32.

Harmon, M. W., Greenberg, S. B., and Couch, R. B. (1976). Effect of human nasal secretions on the antiviral activity of human fibroblast and leukocyte interferon. *Proc. Soc. Exp. Biol. Med.* **152,** 598–602.

——, —— Johnson, P. E., and Couch, R. B. (1977). A human nasal epithelial cell culture system: evaluation of the response to human interferons. *Infect. Immun.* **16,** 480–5.

Hayden, F. G., Mills, S. E., and Johns, M. E. (1983). Human tolerance and histopathologic effects of long term administration of intranasal interferon-alpha 2. *J. Infect. Dis.* **148,** 914–21.

Herzog, Ch., Just, M., Berger, R., Havas, L., and Fernex, M. (1983). Intranasal interferon for contact prophylaxis against common cold in families. *Lancet* **ii,** 962.

Higgins, P. G., Phillpotts, R. J., Scott, G. M., Wallace, J., Bernhardt, L. L., and Tyrrell, D. A. J. (1983). Intranasal interferon as protection against experimental respiratory coronavirus infection in volunteers. *Antimicrob. Agents Chemother.* **24,** 713–5.

Hirsch, M. S., Schooley, R. T., Cosimi, A. B., Russell, P. S., Delmonico, F. L., Tolkoff-Rubin, N. E., Herrin, J. T., Cantell, K., Farrell, M.-L., Rota, T. R., and Rubin, R. H. (1983). Effects of interferon-alpha on cytomegalovirus reactivation syndromes in renal-transplant recipients. *N. Engl. J. Med.* **308,** 1489–93.

Howie, V., Pollard, R. B., Kleyn, K., Lawrence, B., Peskuric, T., Paucker, K., and Baron, S. (1982). Presence of interferon during bacterial otitis media. *J. Infect. Dis.* **145,** 811–4.

Isaacs, D., Clarke, J. R., Tyrrell, D. A. J., Webster, A. D. B., and Valman, H. B. (1981). Deficient production of leucocyte interferon (interferon-α) *in vitro* and *in vivo* in children with recurrent respiratory tract infections. *Lancet* **ii,** 950–2.

Jahiel, R. I., Vilček, J., Nussenzweig, R. S., and Vanderberg, J. (1968). Interferon inducers protect mice against *Plasmodium berghei* malaria. *Science* **161,** 802–4.

Jones, B. R., Coster, D. J., Falcon, M. G., and Cantell, K. (1976). Topical therapy of ulcerative herpetic keratitis with human interferon. *Lancet* **ii,** 128.

Kramer, P., ten Kate, F. W. Y., Bijnen, A. B., Jeekel, J., and Weimar, W. (1984). Recombinant leucocyte interferon A induces steroid-resistant acute vascular rejection episodes in renal transplant recipients. *Lancet* **i,** 989–90.
Levin, S., and Hahn, T. (1982). The interferon system in acute viral hepatitis. *Lancet* **i,** 592–4.
Levis, S., Saaredra, M. C., Ceccoli, C., Falcoff, E., Feuillade, M. R., Enria, D. A., Maiztegui, J. I., and Falcoff, R. (1984). Endogenous interferon in Argentinian hemorrhagic fever. *J. Infect. Dis.* **149,** 428–33.
MacGregor, R. R., Sheagren, J. N., and Wolff, S. M. (1969). Endotoxin-induced modification of *Plasmodium berghei* infection in mice. *J. Immunol.* **102,** 131–9.
McIntosh, K. (1978). Interferon in nasal secretions from infants with viral respiratory infections. *J. Paediatr.* **93,** 33–6.
Merigan, T. C., Gallagher, J. G., Pollard, R. B., and Arvin, A. M. (1981). Short-course human leukocyte interferon in treatment of herpes zoster in patients with cancer. *Antimicrob. Agents Chemother.* **19,** 193–5.
—— Rand, K. H., Pollard, R. B., Abdallah, P. S., Jordan, G. W., and Fried, R. P. (1978). Human leukocyte interferon for the treatment of herpes zoster in patients with cancer. *N. Engl. J. Med.* **298,** 981–7.
—— Reed, S. E., Hall, T. S., and Tyrrell, D. A. J. (1973). Inhibition of respiratory virus infection by locally applied interferon. *Lancet* **i,** 563–7.
Meyers, J. D., Day, L. M., Lum, L. G., and Sullivan, K. M. (1983). Recombinant leukocyte A interferon for the treatment of serious viral infections after marrow transplant: a phase 1 study. *J. Infect. Dis* **148,** 551–6.
—— McGuffin, R. W., Bryson, Y. J., Cantell, K., and Thomas, E. D. (1982). Treatment of cytomegalovirus pneumonia after marrow transplant with combined vidarabine and human leukocyte interferon. *J. Infect. Dis.* **146,** 80–4.
——, —— Neiman, P. E., Singer, J. W., and Thomas, E. D. (1980). Toxicity and efficacy of human leukocyte interferon for treatment of cytomegalovirus pneumonia after marrow transplantation. *J. Infect. Dis.* **141,** 555–62.
Murray, H. W., Rubin, B. Y., Masur, H., and Roberts, R. B. (1984). Impaired production of lymphokine and immune (gamma) interferon in the acquired immunodeficiency syndrome. *N. Engl. J. Med.* **310,** 883–9.
Neumann-Haefelin, D., Sundmacher, R., Skoda, R., and Cantell, K. (1977). Comparative evaluation of human leucocyte and fibroblast interferon in the prevention of herpes simplex virus keratitis in a monkey model. *Infect. Immun.* **17,** 468–70.
Nogueira, N., Kaplan, G., Levy, E., Sarno, E. N., Kushner, P., Granelli-Piperno, A., Colomer-Gould, V., Levis, W., and Steinman, R. (1983). Defective gamma interferon production in leprosy. Reversal with antigen and interleukin 2. *J. Exp. Med.* **158,** 2165–70.
Onwubalili, J. K. and Scott, G. M. (1983). Natural killing and interferon-alpha production in tuberculosis. *Antiviral Res. Abst.* **1,** 75.
Pazin, G., Armstrong, J., Lam, M., Tarr, G., Jannetta, P., and Ho, M. (1979). Prevention of reactivated herpes simplex infection by human leukocyte interferon after operation on the trigeminal root. *N. Engl. J. Med.* **301,** 225–30.
Petralli, J. K., Merigan, T. C., and Wilbur, J. R. (1965). Action of endogenous interferon against vaccinia virus in children. *Lancet* **ii,** 401–5.
Phillpotts, R. J., Scott, G. M., Higgins, P. G., Wallace, J., Tyrrell, D. A. J., and Gauci, C. L. (1983). An effective dosage regimens for prophylaxis against rhinovirus infection by intranasal administration of Hu IFN-α_2. *Antiviral Res.* **3,** 121–36.
Rhodes-Feuillette, A., Druilhe, P., Canivet, M., Gentilini, M., and Périés, J. (1981). Présence d'interféron circulant dans le sérum de malades infectés par *Plasmodium falciparum. C. R. Hebd. Séanc. Acad. Sci. Paris* **293,** 635–7.
Romano, A., Revel, M., Gurari-Rotman, D., Blumenthal, M., and Stein, R. (1980). Use of human fibroblast-derived (beta) interferon in the treatment of epidemic adenovirus keratoconjunctivitis. *J. Interferon Res.* **1,** 95–100.

Sauvaget, F. and Fauconnier, B. (1978). The protective effect of endogenous interferon in mouse malaria as demonstrated by the use of anti-interferon globulins. *Biomedicine* **29,** 184–7.

Schattner, A., Meshorer, A., and Wallach, D. Interferon is involved in virus-induced lymphopenia. *Antiviral Res. Abst.* **1,**(2) 84.

Schellekens, H., Weimar, W., Cantell, K., and Stitz, L. (1979). Antiviral effect of interferon *in vivo* may be mediated by the host. *Nature* **278,** 742–3.

Schonfeld, A., Nitke, S., Schattner, A., Wallach, D., Crespi, M., Hahn, T., Levani, H., Yarden, O., Shoham, J., Doerner, T., and Revel, M. (1984). Intramuscular human interferon-β injections in treatment of condylomata acuminata. *Lancet* **i,** 1038–42.

Schultz, W. W., Huang, K. Y., and Gordon, F. B. (1968). Role of interferon in experimental mouse malaria. *Nature* **220,** 709–10.

Scott, G. M. (1983*a*). The toxicity of interferon in man. In: *Interferon 5* (ed. I. Gresser) pp. 87–114. Academic Press, London.

—— (1983*b*). The antiviral effects of interferon. In: *Interferons: from molecular biology to clinical application. SGM Symposium 35* (ed. D. C. Burke and A. G. Morris) pp. 277–311. Cambridge University Press, Cambridge.

Scullard, R. H., Pollard, R. B., Smith. J. L., Sacks, S. L., Gregory, P. G., Robinson, W. S., and Merigan, T. C. (1981*a*). Antiviral treatment of chronic hepatitis B virus infection: 1. Changes in viral markers with interferon combined with adenine arabinoside. *J. Infect. Dis.* **143,** 772–83.

—— Andres, L. L., Greenberg, H. B., Smith, J. L., Sawhney, V. K., Neal, E. A., Mahal, A. S., Popper, H., Merigan, T. C., Robinson, W. S., and Gregory, P. B. (1981*b*). Antiviral treatment of chronic hepatitis B virus infection: improvement in liver disease with interferon and adenine arabinoside. *Hepatology* **1,** 228–32.

Spruance, S. L., Green, J. A., Chiu, G., Teh, T. J., Wenerstrom, G., and Overall, J. C. (1982). Pathogenesis of herpes simplex labialis: correlation of vesicle fluid interferon with lesion age and virus titre. *Infect. Immun.* **36,** 907–10.

Stevens, D. A. and Merigan, T. C. (1972). Interferon, antibody and other host factors in herpes zoster. *J. Clin. Invest.* **51,** 1170–8.

Sundmacher, R. (1982). Interferon in ocular viral diseases. In *Interferon 4* (ed. I. Gresser) pp. 177–200. Academic Press, London.

Sundmacher, R., Neumann-Haefelin, D., and Cantell, K. (1976). Interferon treatment of dendritic keratitis. *Lancet* **i,** 1406–7.

—— Wigand, R., and Cantell, K. (1980). The value of exogenous interferon in adenovirus keratoconjunctivitis. Preliminary results. *Albrecht von Graefes Arch. Clin. Exp. Ophthalmol.* **218,** 139–40.

—— Cantell, K. and Mattes, A. (1984). Combination therapy for dendritic keratitis with high-titer alpha-interferon and trifluorothymidine. *Arch. Ophthalmol.* **102,** 554–5.

—— Cantell, K., Skoda, R., Hallermann, C., and Neumann-Haefelin, D. (1978). Human leukocyte and fibroblast interferon in a combination therapy of dendritic keratitis. *Albrecht von Graefes Arch. Klin. Exp. Opthalmol.* **208,** 229–33.

Vesterinen, E., Meyer, B., Purola, E., and Cantell, K. (1984). Treatment of vaginal flat condyloma with interferon cream. *Lancet* **i,** 157.

Virelizier, J. L., Lipinski, M., Tursz, T., and Griscell, C. (1979). Defects of immune interferon secretion and natural killer cell activity in patients with immunological disorders. *Lancet* **ii,** 696–7.

Wade, J. C., Hintz, M., McGuffin, R. W., Springmeyer, S. C., Connor, J. P., and Meyers, J. D. (1982). Treatment of cytomegalovirus pneumonia with high dose acyclovir. *Am. J. Med.* **73** (Suppl.) 249–56.

Weimar, W., Stitz, L., Billiau, A., Cantell, K., and Schellekens, H. (1980). Prevention of vaccinia lesions in rhesus monkeys by human leucocyte and fibroblast interfer-

ons. *J. Gen. Virol.* **48,** 25–30.
—— Schellekens, H., Lameijer, L. D. F., Masurel, N., Edy, V. G., Billiau, A., and De Somer, P. (1978). Double blind study of interferon administration in renal transplant recipients. *Eur. J. Clin. Invest.* **8,** 255–8.
Wheelock, E. F. (1964). Interferon in dermal crusts of human vaccinia virus vaccinations: a possible explanation of relative benignity of variolation smallpox. *Proc. Soc. Exp. Biol. Med.* **117,** 650–3.

7 Interferons and maligant disease

A. Nethersell and K. Sikora

7.1. INTRODUCTION

Three out of every ten persons in the world will develop cancer. Improved surgery and more sophisticated radiation techniques have led to improved local control. We now have a wide range of cytotoxic drugs for use in generalized disease. Despite this, the overall outlook for most solid tumours, with a few notable exceptions, has improved little over the last 20 years. There are several reasons for this. Many patients' disease is too widespread at presentation. Chemotherapy for disseminated disease often fails, leading to little real survival benefit. Occult dissemination at presentation invariably presents itself later. Adjuvant systemic treatment aimed at micrometastatic disease has so far conferred no real survival benefit overall, although it has increased the relapse-free interval in some cases. Truly local disease itself may be troublesome with up to 30 per cent dying with uncontrolled disease at the primary site. Reasons for this include radioresistance and incomplete surgery. Furthermore, an ageing population is more prone to cancer and at the same time less able to withstand the rigours of aggressive therapy. It is small wonder, therefore, that a different type of anti-cancer drug should have caught the imagination of press, public, and the medical world. In some quarters this new and unknown agent was exalted to the role of a panacea for cancer. It had taken 20 years from Isaacs and Lindenmann's perceptive and far reaching observations on the resistance to subsequent viral infection of monolayers of cells treated with soluble proteins derived from virus-infected cell cultures, to get adequate quantities of the drug for clinical trial. The last two years have been tremendous activity in the clinical investigation of this drug. We can now evaluate realistically its efficacy as an antitumour agent.

7.2. THE INTERFERONS

The interferon system consists of three types of functionally related proteins or glycoproteins each consisting of a single polypeptide chain of molecular weight 16–26 000 daltons. The three types arise from different cells in response to various inducing agents (Table 7.1). Each type differs antigenically and structurally as well as containing further subtypes, some of whose structures have been completely elucidated. (See Chapter 1 this

Table 7.1. *Types of human interferon used in the treatment of malignant disease*

Type	Whether glycosylated	Usual source	Inducing agents	No. of subtypes to date
IFN-α (leukocyte)	No	Macrophages Lymphoblastoid cells	Viruses, virus-infected cells, tumour cells, bacteria, mitogens	At least 13
IFN-β (fibroblast)	Yes	Epithelial cells, fibroblasts	Viruses, foreign nucleic acids	One definite, perhaps more
IFN-γ (immune)	Yes	T-lymphocytes which have been pre-sensitized to antigens	Foreign antigens and mitogens	One

volume). IFN-γ may well have greater anti-tumour than anti-viral activity as compared to the other interferons, but problems with its stability have until recently hindered its clinical use.

Several sources of interferon have been exploited for clinical use against cancer and it was estimated recently that more than 30 different firms were producing natural interferons by tissue culture techniques or recombinant species by genetic engineering and DNA technology. Three cell types are in general used as a source of IFN.

7.2.1. Normal leukocytes

Here the induced cells are leukocytes from normal donor blood, or rarely from patients with chronic myeloid leukaemia. Cantell, director of the Finnish Red Cross, produced partially purified crude interferon in this way from buffy coat or leukapheresed white cells and this was used in many of the early clinical studies. It should be noted that this interferon was originally less than 1 per cent pure.

7.2.2. Lymphoblastoid cell lines

Most of these are derived from Burkitt's lymphoma and contain the EBV genome, although the commonly used Namalwa line has lost 50 per cent of this genome. Large-scale production is possible and several centres are using these lines to produce interferons containing α:β species in proportions of 5–10:1. The Wellcome preparation (Wellferon from Namalwa) retains only the α species, however, but contains at least eight different subtypes formulated in tris–glycine buffered normal saline with human albumin as stabilizer.

7.2.3. Fibroblasts

Diploid fibroblasts from foreskins or embryos are grown in monolayers. Induction is usually with poly I-rC and cycloheximide and results in β interferon production.

In each of the above methods the supernants contain interferon of only low activity. Various methods of purification are used including potassium thiocyanate and ethanol precipitation, high performance liquid chromatography and two-dimensional polyacrylamide gel electrophoresis, and more recently immuno-absorption by specific monoclonal antibodies. Purity is now high and specific activities are now in the order of 2×10^8 i.u. mg^{-1}.

7.2.4. Recombinant DNA technology

At least two types of IFN-α of almost identical structure have now been produced commercially and will probably soon be available for general use (r IFN-αA, or Roferon A from Hoffmann-La Roche and IFN-α_2, or Intron A from Schering-Plough). Other recombinant interferons have been produced on a smaller scale. These techniques have led to increased knowledge about the interferon α genes themselves; at least 13 seem to be functional while six are non-functional pseudogenes containing stop codons which prevent synthesis of whole proteins. These may represent evolutionary remnants or new genes in evolution.

The techniques dealing with the isolation of DNA molecules coding for interferons and the expression of these genes in plasmic vectors in bacteria and yeast are described in detail in other chapters in this book. Briefly, human cells are grown in tissue culture and challenged with inducing agents to produce interferon mRNA. Cytoplasmic mRNA is extracted from the induced cells. cDNA, complementary to this RNA, can then be produced *in vitro* using reverse transcriptase. Suitable fragments of DNA are then incorporated into plasmid vectors. The DNA carrying the genetic information for interferon production is then ligated into the plasmid circle using DNA ligase and the new plasmid vector reintroduced into the *E. coli* host. Intracellular plasmids enjoy an independent existence and are able to replicate and direct protein synthesis. Suitable colonies of *E. coli* can then be grown in fermentation vats and vast quantities of interferon produced and refined, to achieve 97 per cent purity and a specific activity of 3×10^8 i.u. mg.$^{-1}$

The α-IFNs produced by genes expressed in *E. coli* are biochemical replicas of the corresponding naturally produced molecules, but as *E. coli* cannot glycosylate proteins, IFN-β and-γ lack the carbohydrate side-chains found on the corresponding interferons produced by human cells. Whether this difference will be relevant clinically is not yet known but attempts are now under way to utilize more complex hosts such as yeasts in which glycosylation is possible. Further hopes for the future are that genetically engineered analogues of IFNs may be more specific; a new 'consensus' or average IFN gene has already been synthesized and expressed in bacteria, and other analogues or hybrid molecules are also being developed.

7.3. PRECLINICAL STUDIES

The first animal studies were performed in the 1960s by Dr Ion Gresser in France using interferon of low specific activity. Growth inhibition was seen in several virally induced tumours. It had seemed a reasonable hypothesis that interferon, a known anti-viral agent, might have an effect upon virally induced tumours, but at this stage it was not clear whether the interferon was affecting the tumour or the virus. Subsequent work, largely with mice, showed inhibition of transplantable animal tumours of widely different aetiologies including spontaneous tumours and those induced by viruses and chemical carcinogens. Interferon seemed more effective for those that were virally induced however. Concurrently, purification methods for animal and human interferon were improving rapidly, resulting in larger amounts of purer interferon being available for studies.

During this period *in vitro* studies showed growth inhibition by interferon of a whole range of normal and tumour cells in culture (e.g. lymphoid, leukaemic, osteosarcoma, and breast lines). Cell division was inhibited and furthermore it seemed that tumour cells were more susceptible than their normal counterparts. Cell killing was related to the concentration of interferon and duration of exposure.

The animal data have been summarized by Gresser (1977, 1983) and the following conclusions can be drawn for transplanted animal tumours.

1. Interferon given prior to tumour inoculation is usually ineffective.

2. Limited quantities are usually ineffective; repeated injection is most effective in preventing tumour growth.

3. A *small* tumour burden is more susceptible to interferon than a large one.

4. Tumour inhibition may be related to the total dose per injection as well as to the total cumulative dose per week although the experimental data are not entirely consistent.

5. There is a general feeling that maximal effects are obtained when direct contact between interferon and tumour cells is maximal.

6. Regression of established tumours has rarely been reported although inhibition of growth following inoculation is well authenticated.

7. Inhibition by IFN-γ may well be greater than with IFN-α, but further data are needed.

8. Combinations with cytotoxic agents have in some cases been synergistic rather than purely additive.

These conclusions are important and may have relevance for future clinical work.

7.4. PHARMACOKINETICS AND DOSE RESPONSE DATA

Several phase I studies have investigated optimal dosage schedules. Unfortunately, no broad agreement has yet been reached. No correlation of response with dosage in the range $3–118 \times 10^6$ i.u., i.m. three times weekly, was seen with Roferon A for a range of different tumours (Sherwin, Knost, Fein, Abrams, Foon, Ochs, Schoenberger, Maluish, and Oldham 1982). Tumour responses usually began within two weeks. Peak serum IFN levels after i.m. injection occurred within 4–6 h and increased with increasing dose. The half-life was 6–9 h but for doses greater than 50×10^6 i.u. there was also some steady-state accumulation after 24 h. Similar results have been obtained for lymphoblastoid IFN-α (Wellferon) in a dose-escalation study, and a linear relationship between maximum IFN concentration and dose was inferred in the range $2–30 \times 10^6$ i.u.$^{-2}$, although the scatter of data was considerable. Doses above 15×10^6 i.u. m^{-2} three times weekly were poorly tolerated with changes in higher function as well as systemic disturbance (Laszlo, Huang, Brenckman, Jeffs, Koren, Cianciolo, Metzgar, Cashdollar, Cox, Buckley, Tso, and Lucas 1983). In the Cambridge study with Roferon A, 10×10^6 i.u. m^{-2} daily (or 25×10^6 i.u. m^{-2} three times per week) was tolerated without major CNS disturbance.

Studies so far have shown that the maximum tolerated dose is in the range $40–70 \times 10^6$ i.u. i.m. per week in divided doses. Some studies have used far higher doses than this without undue toxicity whilst others consider that no more than 3×10^6 i.u. day^{-1} should be used for ambulatory patients. Optimal therapeutic schedules have not yet been defined and it seems likely that these will be determined empirically. Three times weekly regimens have the advantage of fewer injections, easier home administration, and an extra day at weekends to recover from side effects.

The use of continuous intravenous infusion of lymphoblastoid and recombinant IFN-α_2 has been investigated (Rohatiner, Balkwill, Griffin, Malpas, and Lister 1982). Escalating doses to 200×10^6 i.u. m^{-2} daily resulted in severe toxicity with drowsiness, disorientation and grand mal fitting at the upper level. Hypotensive episodes were also seen at lower dosage. Nonetheless, 100×10^6 i.u. m^{-2} in 24 h was better tolerated than expected and was considered the maximal permissible dose. Peak serum levels were usually reached after 24 h of infusion, but as with i.m. studies there was considerable patient variation. No urinary excretion of IFN was found and CSF levels were unmeasurable in four of five patients and present at low concentration in the fifth. IFN does not cross the blood–brain barrier.

7.5. TOXICITY

Impurities in the early IFN preparations were originally thought to be responsible for most of the toxic side effects. Nearly all of these have persisted, however, and seem to be common to all species so far used, despite 97 per cent purification.

7.5.1. Fever

This invariably appears with doses above 1×10^6 i.u. m^{-2}, appearing a few hours after IFN administration and usually settling within one week despite continued IFN administration. Following a gap in IFN therapy pyrexia usually reappears upon recommencing treatment. Antipyretics and steroids are effective in preventing it. The cause is probably central in origin, possibly mediated via prostaglandins which are known to be stimulated by interferons.

7.5.2. General

Fatigue, lethargy and anorexia are invariably seen in varying degrees, along with arthralgia, myalgia, chills, and malaise resembling a 'flu-like' syndrome. Worsening fatigue and lethargy tend to precede major CNS toxicity.

7.5.3. Nervous system

Peripheral effects include paraesthesia and numbness; central effects include confusion, disorientation, dysphasia, seizures, and coma. These appear to be dose-related and are usually reversible. The features are reminiscent of a metabolic encephalopathy (Smedley, Katrak, Sikora, and Wheeler 1982). EEGs generally show the development of diffuse slow-wave activity. Minor EEG changes precede those seen during major CNS toxicity and may resolve only slowly after cessation of therapy. The cause of these changes is unknown.

7.5.4. Weight loss

This may be considerable and is the result of anorexia, nausea, vomiting, GI disturbance, lethargy, and lack of volition.

7.5.5. Cardiac

UK and American studies have seen few cardiac complications. Two myocardial infarctions with arrhythmias occurred in France after the administration of a relatively impure IFN preparation.

7.5.6. Haematological

Leukocyte and platelet counts almost always fall markedly though rarely to dangerous levels. Haemoglobin fall is usually less marked and may slowly right itself. Reticulocyte response is impaired and cultured bone marrow shows reduced colony growth, suggesting a direct effect of interferon upon erythropoiesis and granulopoiesis. Margination of leukocytes may also occur.

7.5.7. Hepatic

Liver lactate dehydrogenase and transaminase levels usually show moderate rises by the third week and then return to normal. Occasionally very high values occur which are usually dose-related and suggestive of hepatic damage. Bilirubin and alkaline phosphatase are little affected and hepatic side effects do not correlate with encephalopathy.

7.5.8. Other effects

Alopecia, dry mouth, malar flush and sore throat have all been recorded. Electrolyte disturbances such as hyperkalaemia and hypocalcaemia have been seen.

Major side effects are probably dose-related and are by no means uncommon. There is no evidence as yet that any one interferon preparation is less toxic than others. Despite toxicity several have commented on the reluctance of patients to discontinue interferon. In some cases this seems almost more than a psychological dependence.

7.6. CLINICAL STUDIES

7.6.1. Ostecsarcoma

Interferon was first used in cancer therapy at the Karolinska Hospital, Stockholm in 1972 (Strander, Aparisi, Blomgren, Bröstrom, Cantell, Einhorn, Ingimarsson, Milsonne, and Söderberg 1982). Fifty-one patients treated by amputation or radical resection for localized osteosarcoma were subsequently entered into the study and given 3×10^6 i.u. of Cantell IFN daily and subsequently three times weekly for up to one and a half years as adjuvant treatment. Previous laboratory work had shown dose-related growth inhibition of osteosarcoma cell lines. Owing to the rarity of the condition and poor outlook, no controls were included in the study. The treatment group was therefore compared with two controls, one of them historical, containing 35 patients, and the second a concurrent group of 51 cases treated at other Swedish centres. Twenty-one of these also received high dose adjuvant cytotoxic therapy. The results are summarized in Table 7.2.

Table 7.2. *Results of adjuvant interferon therapy in osteosarcoma*

	No. of cases	Five year survival (per cent)	Per cent disease-free at five years
IFN group	51	54	44
Historical group	35	14	14
Concurrent Swedish group	51	36	33

Interpretation of these results is difficult because prognostic factors, such as histological grading, were worse in the historical group, suggesting that the overall aggressiveness of the disease may have lessened over the years, a hypothesis supported by improved results following surgery alone at the Mayo Clinic. The difference between the IFN and concurrent groups, though not statistically significant, is encouraging. Furthermore, 21 of the concurrent group were given chemotherapy and suffered moderately toxic side effects. The IFN group seem, on the other hand, to have tolerated their treatment well. More conservative surgery was also used in the IFN group (wide resection rather than amputation) without any increase in failure rate.

Eight patients also received concurrent IFN and pulmonary irradiation (2000 cGy maximum) for lung metastases. All improved and there was radiological regression in two. One might, perhaps, have hoped for better than this if IFN and radiation are at all synergistic as has been suggested (Valter 1977; Dritschilo, Mossman, Gray, and Sreevalsan 1982).

7.6.2. Leukaemia

Several laboratory studies have suggested that IFN might be useful in leukaemia: e.g. inhibition of mouse leukaemia cells both in culture and after transplantation as well as a reduced leukaemia incidence and increased survival in IFN-treated AKR mice (mice which normally develop spontaneous leukaemia at six months). Other studies have been directed at human cell culture systems (Balkwill and Oliver 1977) and for myeloblasts in particular, 50 per cent growth inhibition can be obtained at a concentration of 10^3 i.u. ml^{-1} of α-interferon. This level can be achieved clinically by continuous infusion of doses greater than 30×10^6 i.u. m^{-2} day.$^{-1}$.

The most extensive clinical studies have taken place at St Bartholomew's Hospital in London using both lymphoblastoid interferon and IFN-α_2 (Rohatiner *et al.* 1982). Results of phase I studies have sadly been disappointing.

In acute lymphoblastic leukaemia (ALL) the peripheral blast count usually falls with interferon without any change in marrow infiltration. In the acute myeloid leukaemia (AML) study, three of six patients showed no response, two showed a fall in peripheral blasts, but only one showed complete clearing of peripheral blasts and normalization of the bone marrow. Unfortunately, this patient also had a soft tissue deposit of AML which failed to regress completely and so complete remission with IFN could not be claimed. The importance of this study includes the definition of maximum tolerable continuous intravenous dose (10^8 i.u. m^{-2} 24 h^{-1}) and the discrepancy between the observed clinical response and the *in vitro* studies in which growth inhibition by interferon of patients' myeloid cells in culture was seen despite the negative clinical response.

In chronic lymphocytic leukaemia (CLL) response rates similar to those for grade I lymphomas are to be expected and current interest includes combinations of interferon with cytoxic drugs. In chronic myeloid leukaemia (CML) reduction in the circulating white count has been achieved and in one study maintenance interferon therapy seemed able to keep the count under control.

More recently, remarkable results have been seen in hairy-cell leukaemia

(Quesada, Reuben, Manning, Hersh, and Gutterman 1984), a slowly relentless disease which rarely responds to chemotherapy, and usually only transiently to splenectomy. Seven patients with progressive disease were given 3×10^6 i.u. of partially purified leukocyte IFN daily. Three attained complete bone marrow remission and three showed less than 5 per cent leukaemic cells in the aspirate. Haemoglobin level, white cell, and platelet counts returned to normal in all cases in which they were originally deranged. Remissions lasted from over six to over 10 months.

7.6.3. Lymphoma

Several lymphoid cell lines, including Burkitt's lymphoma, show growth inhibition at low IFN concentrations. Human–mouse hybridomas produced by fusing neoplastic human lymphoid cells with mouse cells also showed dose-dependent growth inhibition (Sikora, Basham, Dilley, Levy, and Merigan 1980). Spontaneous lymphomas in susceptible strains of mice appeared less frequently in those that were treated with interferon and their life span was also increased.

Several phase I studies have shown partial or less commonly complete responses in up to 50 per cent of patients with a variety of lymphoid malignancies (Merigan, Sikora, Breeden, Levy, and Rosenberg 1978). The most marked responses were seen in nodular lymphomas and CLL. Few responses occurred in patients with histiocytic lymphomas. Nodular lymphomas often have less aggressive time courses with occasional remissions or waxing and waning of the disease (Krikorian, Portlock, Cooney, and Rosenberg 1980). It is tempting to postulate that IFN is acting by immunological modulation in these cases rather than directly as an antineoplastic agent.

Phase II studies arranged by the American Cancer Society (Horning, 1983) have confirmed the earlier data and shown partial responses in 30 per cent of lymphomas of favourable histology. Toxicity appears not to have been a major problem with Cantell IFN at dosages of up to 9×10^6 i.u. daily, which were generally necessary to obtain any sort of response. Stabilization of disease has also been seen in patients failing to show definite regression. Further phase II and III studies are in progress or planned and interest has been shown in combining interferon with chemotherapy regimes, since it does not compare favourably with standard cytotoxic drugs as a single agent. In the future, specific subtypes of interferon may be found which regulate lymphoid division, and these may also be active against the various lymphomas. It is also hoped that synergism may exist between interferon and cytotoxic drugs as suggested by *in vitro* and animal studies (Chirigos and Pearson 1973; Balkwill and Moodie 1984).

7.6.4. Myeloma

Several phase II studies have shown the interferons to have some effect in myeloma. In Scandinavia, standard treatment consisting of melphalan plus prednisolone has been compared with interferon as a single agent in 43 patients. Six out of 22 melphalan patients responded and four out of 22 interferon patients, although survival was less than in the interferon arm (Mellstedt, Ahre, Bjorkholm, Johansson, Strander, Brenning, Engstedt, Gahrton, Holm, Lehrner, Longvist, Nordenskjold, Killander, Stalfeldt, Simonsson, Ternstedt, and Wadman 1982). Similar results were obtained in the M. D. Anderson study with patients previously untreated by chemotherapy appearing to fare better, and remissions lasting three to 33 months (Alexanian, Gutterman, and Levy 1982). Other studies have confirmed a response rate of about 25 per cent or higher, even for patients refractory to chemotherapy. In the UK however, few responses have been seen with lymphoblastoid IFN and IFN-β also seemed to have little effect (Priestman 1982).

Myeloma should prove to be an ideal neoplasm for assessment since the tumour-secreted monoclonal immunoglobulin can be measured precisely and used as a disease marker. The overall response rate of perhaps 20 per cent is encouraging for a disease which has a poor prognosis and for which cytotoxic drugs provide only temporary remissions. Recombinant α-interferons will soon be available for phase III trials comparing chemotherapy with and without interferon. There is also some anecdotal evidence that interferon priming may resensitize myeloma to chemotherapy to which it has previously become refractory (Hager 1983).

7.6.5. Melanoma

Melanoma is an interesting neoplasm in that immunological factors in the host may modify the course of the disease. Evidence for this is the profuse lymphocytic infiltration sometimes seen in secondarily involved lymph nodes, and spontaneous remission or stabilization of disease at certain sites. Phase I studies which included the use of recombinant IFN-α showed partial responses.

A recent phase II study with Wellferon showed only one partial response in 17 patients with advanced metastatic disease, 16 of whom had received prior chemotherapy (Retsas, Priestman, Newton, and Westbury 1983). All patients had progressive disease at entry although six had previously been stabilized with chemotherapy. The initial dose was 2.5×10^6 i.u. m^{-2} i.m. daily in most cases but dose reductions or cessation of treatment were often necessary because of toxicity. Interestingly, three patients who were receiving dexamethasone for CNS complications did not develop the expected fever after i.m. injection, although one of these became pyrexial when the drug was given intravenously. One patient showed definite histological regression of most cutaneous melanoma nodules leaving only residual pigmentation in the skin. Eight months after starting interferon new nodules appeared at other sites although the patient was still alive with only minimal disease after 13 months of treatment. It is perhaps pertinent, however, that the metastases involved only the skin. As with other tumours, regression at one site with progression elsewhere was observed in some cases.

Despite the limited activity of lymphoblastoid interferon in this particular study several American investigators have seen significant responses in up to a third of the patients. Retrospective analysis of some of these studies has shown 10 out of 39 responses. More recently, out of six patients given IFN-α_2, two responded (one of these completely) and two showed stabilization of disease (Ernstoff, Davis, and Kirkwood 1984). With response rates of no more than 20 per cent with toxic chemotherapy regimes any new agent such as this is worthy of further trial.

A most fascinating report describing the combination of interferon with cimetidine in metastatic meloma appeared last year (Borgström, von Eyben, Flodgren, Axelsson, and Sjögren 1982) and further investigation is under way at present. Six patients were treated initially, all of whom had ongoing disease and had failed previously to respond to IFN alone. The combination resulted in two complete remissions, one partial remission and one stabilization of disease. Serendipity played its part, for the discovery resulted from the observation that a patient with melanoma who was showing no response to interferon responded rapidly after starting cimetidine for a gastric ulcer. H2 antagonists are known to inhibit a variety of regulatory functions mediated by T suppressor cells *in vitro* as well as showing anti-tumour activity in animal models, and it is thought that inactivation of suppressor T cells may occur *in vivo*. The role of cimetidine as a single agent is also under review, but preliminary studies have shown tumour progression (Borgström, S., von Eyben, Flodgren, Axelsson, and Sjögren 1983). It seems likely that interferon may be necessary to activate both phagocytic cells and lymphocytes.

7.6.6. Renal carcinoma

Metastatic renal carcinoma is usually resistant to all conventional forms of chemotherapy. Spontaneous regression of metastatic disease (usually pulmonary) though rare has been repeatedly described. Hormone manipulation (usually with progestogens) can also occasionally lead to regression. American Cancer Society studies using leucocyte IFN showed one complete, one partial and several minor responses in 17 patients. A second study showed improvement or stabilization in 16 out of 40 patients. Cantell IFN has also been used in the M. D. Anderson study (Quesada, Swanson, Trinidade, and Gutterman 1983) at a dose of 3×10^6 i.u. i.m. daily (or larger doses twice weekly). Toxicity was not dose limiting and results were encouraging with five achieving partial responses, two less than partial and three with mixed responses (regression at one site and progression elsewhere). Responses were limited to lung or mediastinum and occurred within 30–90 days of starting treatment, lasting from 6–12 months. Response appeared to correlate with previous long disease-free interval (suggesting slowly progressive disease), high Karnofsky status and interferon induced leukopenia. Neither it nor myelosuppression correlated with serum interferon levels. Dose escalation failed to provide any further improvement. These results compare favourably with most chemotherapy schedules.

7.6.7. Lung cancer

New agents are desperately needed in this condition for which survival has only marginally increased recently despite costly and debilitating chemotherapy regimes. Animal studies were not particularly encouraging for this tumour in that growth inhibition of inoculated cells was only achieved with interferon doses 100–1000 times those used clinically. It is, therefore, not surprising that clinical studies have been so disappointing.

For non-small-cell carcinoma no responses were seen in 11 evaluable patients given 3×10^6 i.u. daily i.m. For small-cell carcinoma (Jones, Bleehen, Slater, George, Walker, and Dixon 1983) 10 previously untreated patients received $50–100 \times 10^6$ i.u. m^{-2} of lymphoblastoid interferon i.v. by continuous infusion for five days followed by 3×10^6 i.u. u.m. three times weekly for three weeks. No responses were seen and six patients showed progression of disease during this time. Toxicity was considerable and treatment abandoned in two patients. Three patients with inappropriate ADH secretion and hyponatraemia showed marked deterioration during interferon therapy, a side effect not previously described. Calcium levels sometimes fell during the infusion.

A similar study in Helsinki (Mattson, Niiranen, Holsti, Iivanainen, Bergström, Standertskjöld-Nordenstam, Tarkkanen, Färkkilä, Anderson, Hilsti, Kauppinen, and Cantell 1982), using a higher maintenance dose for longer claimed three out of eight minor responses after 5–20 weeks of treatment. It was felt that interferon might have some growth-delaying effect as survivals were longer than expected, although the number of patients was only small. The Helsinki study would suggest that interferon may have minimal activity in this condition.

The possibility that interferon and cytotoxic drugs or radiation may be synergistic has previously been mentioned and there is some laboratory evidence for this (Dritschilo, Mossman, Gray, and Sreevalson 1982) in that cell survival studies with mouse 3T3 cells which were irradiated with and without prior exposure to interferon showed marked reduction in the shoulder of the cell survival curve for the interferon group. The Helsinki study included six patients who received interferon as well as conventional radiotherapy (5500 cGy in 20 fractions over 7 weeks, split course). Four of these six patients developed radiation pneumonitis which was thought to be unexpectedly severe and possibly related to interferon priming (Mattson and Holsti 1983).

7.6.8. Breast cancer

In contrast to the pre-clinical studies for lung cancer, several reports suggested that interferon might be a useful agent in breast cancer. Spontaneous and xenografted animal tumours as well as tissue culture work with human breast carcinoma lines showed definite evidence for interferon's activity.

The results of phase I and phase II trials to date are conflicting. The data from two phase I trials (American Cancer Society and M. D. Anderson study) which used Cantell interferon in dosage of $3\text{–}9 \times 10^6$ i.u. daily for 28 days, proceeding in some cases to maintenance, have recently been analysed. Using pooled data from these two studies, 11 of 40 assessable patients had partial responses of greater than 50 per cent reduction in the product of maximum diameters of assessable tumour, whilst five had lesser responses. Response duration seemed to be related to continued treatment with median values of 196 days with maintenance IFN and 91 days without. Response was thought to correlate with the degree of leukocyte depression, increasing age and, surprisingly, the development of herpes labialis whilst on interferon (although this was not statistically significant). The true partial response rate for the pooled data was thought to lie between 15 and 38 per cent with a 95 per cent confidence limit.

Nowhere near this response rate has been seen with Roferon A despite the hopeful phase I studies. Two phase II trials have taken place (Sherwin, Mayer, Ochs, Abrams, Knost, Foon, Fein, and Oldham 1983; Nethersell, Smedley, and Sikora 1984). In the first, 17 were evaluable, and 16 of these showed progressive disease. Dose reductions were necessary in all patients. The Cambridge study showed similar results, with toxicity a major problem. The doses of 20×10^6 i.u. m^{-2} daily or 50×10^6 i.u. m^2 three times per week, which had appeared acceptable in phase I studies, were too high and had to be reduced. Fifteen patients with assessable advanced breast cancer, who had failed at least one conventional form of therapy, were entered into the study. At four weeks two out of 12 appeared to have partial responses, and 7 had lesser responses, but by 12 weeks only two patients showed any degree of response at all. Both of these had a history of somewhat indolent disease which, although locally advanced, had been present for a long time and had not metastasized widely. Reduction of metastatic bone pain with a fall in alkaline phosphatase activity was seen twice.

It is difficult to reconcile the above results. Possibly Cantell IFN, which contains up to eight α-IFNs, may be more biologically active or may even contain other active molecules with anti-tumour activity. Furthermore, the role of the interferons as adjuvants or in combination is not yet defined.

7.6.9. Kaposi's sarcoma

Immunological mechanisms almost certainly play a role in the development of Kaposi's sarcoma in patients suffering from the acquired immunodeficiency syndrome (AIDS). The cause of this syndrome, which usually affects young homosexual males, is one of the mysteries of the decade, although persistent cytomegalovirus infection has been suggested. Sufferers show impaired cell-mediated responses and susceptibility to opportunistic infection. Many develop Kaposi's sarcoma involving skin, lymph nodes, and GI tract in variable proportions. Evidence that the tumour is related to a breakdown of immunological integrity is afforded by its occurrence in immunosuppressed transplant patients whose tumours can regress spontaneously when immunosuppression is lifted. Since IFN can modulate the immune response as well as having anti-viral and anti-tumour actions it seems a logical choice as therapeutic agent in these patients who would not easily tolerate conventional cytotoxic chemotherapy.

Krown, Real, Cunningham-Rundles, Myskowski, Koziner, Fein, Mittleman, Oettgen, and Safai (1983) have reported three complete responses, two partial responses and three lesser ones in 12 evaluable patients given Roferon A 36×10^6 i.u. or 54×10^6 i.u. daily for 28 days, and thereafter three times weekly if responding. The dose was reasonably well tolerated (certainly better than in older patients with other malignancies) as only two patients required dose reduction. There was some evidence of a dose response effect as three of the minor responders at one month showed progressive disease when changed to three times weekly schedules. Tests of immune function showed increased NK activity (but this failed to correlate with response), a rise in the lymphocyte proliferative response to phytohaemagglutinin in some cases, but no change in DNCB skin testing. There were no major infective episodes. Clearly the role of the interferons in this condition requires further study.

7.6.10. Carcinoid tumours

Most carcinoid tumours (argentaffinomas) arise in the appendix and are benign; those arising in the small bowel are usually malignant. Vasoactive amines (in particular HIAA) secreted by these tumours are released into the portal circulation and inactivated in the liver. The presence of liver metastases, however, leads to the release of these hormones and enzymes into the systemic circulation where they produce flushing, diarrhoea, and bronchospasm, the triad of the carcinoid syndrome, as well as right-sided valvular disease. Metastatic disease is not particularly sensitive to chemotherapy and often not amenable to surgery or embolization with overall survival of 30–40 per cent at five years, although the pace of the disease is usually slow.

The effect of purified Cantell IFN in these rare tumours has recently been reported (Oberg, Funa, and Alm 1983). Nine patients with progressive disease, seven of whom had previously received streptozotocin and 5-fluorouracil, were given 3×10^6 i.u. IFN daily for a month and subsequently twice this dose for a further two months. All of them had systemic symptoms at the outset and six showed clear evidence of reduction in these symptoms after treatment. Eight patients showed a reduction in urinary HIAA levels which rose again in three of these after stopping treatment. Other marker substances, such as α and β HCG and pancreatic polypeptide levels were affected less dramatically. All of the responding patients had liver metastases (as distinct from lymph node spread) and all of these showed falling markers of one sort or another. No change was seen in tumour size on repeat scanning after three months of IFN.

It is unclear whether interferon was directly cytotoxic in this study (unlikely in view of the rebound phenomenon seen in some cases) or whether its action was one of modulation of synthesis, secretion or even degradation of these active amines and polypeptides. The same investigators are now comparing interferon and chemotherapy for these rare tumours.

7.6.11. Other tumours

Gastrointestinal cancer is almost invariably managed primarily by surgery. Persistence or recurrence after surgery is not usually chemosensitive and carries a uniformly poor prognosis. Liver metastases are, of course, frequent. Preliminary results would suggest little place for IFN in these tumours and in particular, phase II studies with IFN-α_A in colorectal cancer have shown almost universal progression of disease (Figlin, Callaghan, and Sarna 1983).

Bladder papillomas are not considered malignant in the true sense of the word, although they frequently recur and eventually may become truly malignant. Injection of interferon into the bases of recurrent and resistent papillomata has resulted in clinical and histological regression (Ikić, Nola, Maricić, Smud, Oresić, Knezević, Rode, Jusić, and Soos 1981).

Topical or intralesional interferon has been used for several other tumours by the same investigators. For *carcinoma of the cervix* normalization of cytological and histological findings in both invasive and *in situ* lesions has been shown. A 30 per cent cure rate is also claimed for repeated intralesional or topical interferon application for selected *head and neck tumours,* including basal and squamous cell carcinoma.

An interesting Japanese study has looked at systemic or local application (via an Omaya valve) of leukocyte interferon on different intracerebral tumours, largely *gliomas*. Responses were seen in those tumours with low growth rates but rapidly growing tumours showed no response. No survival benefit was claimed (Ueda, Hirakawa, Nakagawa, Suzuki, and Kishida 1983).

In addition to bladder papillomas other superficial *papillomas and warts* have been treated by repeated intralesional injection or application of interferon pastes and ointments to pre-traumatized warty excrescences. Since most warts are considered of viral origin it is not surprising that there have been good results with IFN therapy from several centres. Fibroblast interferon has also resulted in significant reduction in the growth of penile warts in one British study. Warts can rarely undergo malignant change and the flat cervical plaques caused by strains of papilloma virus and thought to predispose of cervical dysplasia and carcinoma in young women may well prove to be an interesting subject for study. Good results have also been obtained with laryngeal papillomata, a condition of multiple warty outgrowths in the larynx of young children. A wide range of interferons have been used. Significant regressions have been seen in some cases and Strander and Cantell showed that repeated i.m. interferon helps to prevent recurrence after surgical removal, in addition to being of value as primary management. Gobel, Arnold, Wahn, Treuner, Jurgens, and Cantell (1983) have also shown response to subcutaneous leukocyte interferon, but complete failure of response to fibroblast interferon, suggesting yet again that particular interferons may be more specific in certain situations.

7.7. CONCLUSION

A review of the recent literature would suggest that to date some response can be hoped for in 20–30 per cent of patients (Table 7.3) with certain

Table 7.3. *Combined published response data on interferon and solid tumours*

		Response		
Disease	No. of patients	CR	PR	RR (%)
Breast	55	0	12	22
Colon	32	0	2	6
Hodgkin's	10	0	0	0
Kaposi's sarcoma	12	3	3	50
Lung	52	0	1	2
Melanoma	56	0	3	5
Myeloma	54	3	13	30
N.H.L. (good)	27	4	3	26
N.H.L. (bad)	12	0	0	0
Osteosarcoma	11	0	0	0
Ovary	5	0	1	20
Prostate	10	1	2	30
Renal	65	1	12	20

CR – complete response.
PR – partial response.
RR – response rate (%).

As at January 1984.

specified tumours. It should be borne in mind that most patients so studied have near end-stage disease and are usually unwell, often old, and frequently have tired or flagging immune systems. Studies continue to identify whether certain interferons may be tumour-specific: gamma-interferons may prove particularly interesting in this respect. To date it would seem that IFNs are most likely to be useful in those conditions in which immunological mechanisms may contribute to tumour regulation (lymphoreticular disorders, Kaposi's sarcoma in AIDS victims).

Various mechanisms may account for interferon's tumoricidal activity (Toy 1983). These include effects on the tumour such as direct inhibition of cell division by slowing the cycling time, which, whilst affecting normal cells as well, may have greater effect in terms of cell loss on a rapidly proliferating tumour system. Interferon may also affect changes in cell metabolism and behaviour following membrane binding and a reversion of the transformed malignant cell to a more normal phenotype (a phenomenon seen *in vitro*). Host responses are also probably important as they are known to be modified by the interferons. Augmentation of the cytotoxicity of NK cells for malignant cells, increased macrophage phagocytosis and changes in cell-mediated and antibody responses may all play a part in tumour regulation. The dose and timing of interferon may be important in this respect, and at least one study has shown that higher NK activity is maintained by using lower doses of interferon – an observation which may be of importance when toxicity is considered (Laszlo, Huang, Brenckman, Jeffs, Koren, Cianiolo, Metzgar, Cashdollar, Cox, Buckley, Tso, and Lucas 1983). IFNs may also affect secretion of a variety of agents (hormones, amines, prostaglandins) which may be important in relation to tumour growth or the establishment of micrometastatic disease, as well as influencing the micro-environment of the tumour.

Future developments will take place on three broad fronts. Molecular biology will provide a greater understanding of altered mechanisms in malignant cells and how the interferons can modify these; immunology will hopefully tell us more of the controlling factors in the malignant process. Concurrently, more interferons will become available (perhaps genetically tailor-made for a particular tumour type) and further clinical studies of these alone or in combination will be necessary for early as well as end-stage disease. Admittedly the honeymoon is over, but fascinating developments may well lie ahead for biological tumour agents like interferons, whose mechanism of action is very different from that of conventional drugs.

7.8. REFERENCES

Alexanian, R., Gutterman, J., and Levy, H. (1982). Interferon treatment for multiple myeloma. *Clin. Haematol.* **II,** 211–20.

Balkwill, F. R. and Moodie, E. M. (1984). Positive interactions between human interferon and cyclophosphamide or adriamycin in a human tumor model system. *Cancer Res.* **44,** 804–908.

—— and Oliver, R. T. O. (1977). Interferons as cell regulatory molecules. *Int. J. Cancer* **20,** 500–5.

Borgström, S., von Eyben, F. E., Flodgren, P., Axelsson, B., and Sjögren, H. O. (1982). Human leukocyte interferon and cimetidine for metastatic melanoma. *N. Engl. J. Med.* **307,** 1808–1.

——, ——, ——, ——, and —— (1983). Combination of cimetidine with other drugs for treatment of cancer. *N. Engl. J. Med.* **308,** 592.

Chirigos, M. A. and Pearson, J. W. (1973). Cure of murine leukaemia with drug and interferon treatment. *J. Nat. Cancer Inst.* **51,** 1367–8.

Dritschilo, A., Mossman, K., Gray, M., and Sreevalsan, T. (1982). Potentiation of radiation injury by interferon. *Am. J. Clin. Oncol.* **5,** 79–82.

Ernstoff, M. S., Davis, C. A., and Kirkwood, J. M. (1984). A phase II trial of human leukocyte interferon in patients with malignant melanoma. *Proc. Am. Soc. Clin. Oncol.* **3,** 62.

Figlin, R., Callaghan, M., and Sarna, G. (1983). Phase II trial of α (human leukocyte) interferon administered daily in adenocarcinoma of the colon/rectum. *Cancer Treat. Rep.* **67,** 493–4.

Gobel, V., Arnold, W., Wahn, V., Treuner, J., Jurgens, H., and Cantell, K. (1981). Comparison of human fibroblast and leukocyte interferon in the treatment of severe laryngeal papillomatosis in children. *Eur. J. Paediat.* **137,** 175–6.

Gresser, I. (1977). Antitumour effects of interferon. In *Cancer – a comprehensive treatise. Chemotherapy,* (ed. F. Becker) Vol. 5, pp. 52–71. Plenum Press, New York.

Gresser, I. (1983). The antitumour effects of interferon in mice. In *Interferon and cancer,* (ed. K. Sikora) pp. 65–76. Plenum Press, New York.

Hager, T. (1983). The interferon–cancer trials: hardly hopeless but not too heartening. *J. Am. Med. Ass.* **250,** 1007–10.

Horning, S. (1983). Lymphoma. In *Interferon and cancer* (ed. K. Sikora) pp. 77–83. Plenum Press, New York.

Ikić, D., Nola, P. Maricić, Z., Smud, K., Oresić, V., Knezević, M., Rode, B., Jusić, D., and Soos, E. (1981). Application of human leukocyte interferon in patients with urinary bladder papillomatosis, breast cancer, and melanoma. *Lancet,* **i,** 1022–4.

Jones, D. H., Bleehen, N. M., Slater, A. J., George P. J. M., Walker, J. R., and Dixon, A. K. (1983). Human lymphoblastoid interferon in the treatment of small cell lung cancer. *Br. J. Cancer* **47,** 361–6.

Krikorian, J. G., Portlock, C. S., Cooney, P., and Rosenberg, S. A. (1980). Spontaneous regression of non-Hodgkin's lymphoma: a report of nine cases. *Cancer* **46,** 2093–9.

Krown, S., Real, F. X., Cunningham-Rundles, S., Myskowski, P. L., Koziner, B., Fein, S., Mittelman, A., Oettgen, H. F., and Safai, B. (1983). Preliminary observations on the effect of recombinant leukocyte A interferon in homosexual men with Kaposi's sarcoma. *N. Engl. J. Med.* **308,** 1071–6.

Laslo, J., Huang, A. T., Brenckman, W. D., Jeffs, C., Koren, H., Cianciolo, G., Metzgar, R., Cashdollar, W., Cox, E., Buckley III, C. E. Tso, C. Y., and Lucas, V. S. Jr. (1983). Phase I study of pharmacological and immunological effects of human lymphoblastoid interferon given to patients with cancer. *Cancer Res.* **43,** 4458–66.

Mattson, K., Niiranen, A., Holsti, L. R., Iivanainen, M., Bergström, L., Standertskjöld-Nordenstam, C. G., Tarkkanen, J., Färkkilä, M., Anderson, L., Hilsti, P., Kauppinen, H.-L., and Cantell, K. (1982). High-dose human leukocyte interferon given as a five day continuous intravenous infusion followed by low-dose intramuscular maintenance treatment in previously untreated patients with small cell carcinoma of the lung. Preliminary results. *Eur. J. Resp. Dis. Suppl.* **125,** 71.

Mattson, K. and Holsti, L. (1983). Lung cancer. In *Interferon and Cancer* (ed. K. Sikora) pp. 113–9. Plenum Press, New York.

Mellstedt, H., Ahre, A., Bjorkholm, M., Johansson, B., Strander, H., Brenning, G., Engstedt, L., Gahrton, G., Holm, G., Lehrner, R., Longvist, B., Nordenskjold, B., Killander, A., Stalfeldt, A.-M., Simonsson, B., Ternstedt, B., and Wadman, B. (1982). Interferon therapy of patients with myeloma. In *Immunotherapy of human cancer* (ed. W. Terry and S. Rosenberg). pp. 387–95. Elsevier, New York.

Merigan, T. C., Sikora, K., Breeden, J. H., Levy, R., and Rosenberg, S. A. (1978). Preliminary observations on the effect of human leukocyte interferon in non-Hodgkin's lymphoma. *N. Engl. J. Med.* **299,** 1449–53.

Nethersell, A. B. W., Smedley, H. M., and Sikora, K. (1984). Recombinant interferon in advanced breast cancer. *Br. J. Cancer* **49,** 615–20.

Oberg, K., Funa, K., and Alm, G. (1983). Effects of leukocyte interferon on clinical symptoms and hormone levels in patients with mid-gut carcinoid tumours and carcinoid syndrome. *N. Engl. J. Med.* **309,** 129–33.

Priestman, T. (1982). The present status of clinical studies with interferons in cancer in Britain. *Phil. Trans. R. Soc. Lond.* **B 299,** 119–24.

Quesada, J., Swanson, D. A., Trindade, A., and Gutterman, J. U. (1983a). Renal cell carcinoma: antitumour effects of leukocyte interferon. *Cancer Res.* **43,** 940–47.

Quesada, J. R., Reuben, J., Manning, J. T., Hersh, E. M., and Gutterman, J. U. (1984). Alpha interferon for induction of remission in hairy cell leukemia. *N. Engl. J. Med.* **310,** 15–18.

Retsas, S., Priestman, T. J., Newton, K. A., and Westbury, G. (1983). Evaluation of human lymphoblastoid interferon in advanced malignant melanoma. *Cancer 51,* 273–6.

Rohatiner, A., Balkwill, F. R., Griffin, D. B., Malpas, J. S., and Lister, T. A. (1982). A Phase I study of human lymphoblastoid interferon administered by continuous intravenous infusion. *Cancer Chemother. Pharmacol.* **9,** 97–102.

Sherwin, S. A., Knost, J. A., Fein, S., Abrams, P. G., Foon, K. A., Ochs, J. J., Schoenberger, C., Maluish, A. E., and Oldham, R. K. (1982). A multiple-dose Phase I trial of recombinant leukocyte A interferon in cancer patients. *J. Am. Med. Ass.* **248,** 2461–6.

Sherwin, S. A., Mayer, D., Ochs, J. J., Abrams, P. G., Knost, J. A., Foon, K. A., Fein, S., and Oldham, R. K. (1983). Recombinant leukocyte A interferon in advanced breast cancer. *Ann. Int. Med.* **98**, 598–602.

Sikora, K., Basham, T., Dilley, J., Levy, R., and Merigan, T. C. (1980). Inhibition of lymphoma hybrids by human interferon. *Lancet* **ii,** 891–3.

Smedley, H. M., Katrak, M., Sikora, K., and Wheeler, T. (1982). Neurological effects of recombinant human interferon. *Br. Med. J.* **286,** 262–4.

Strander, H., Aparisi, T., Blomgren, H., Bröstrom, L.-A., Cantell, K., Einhorn, S., Ingimarsson, S., Milsonne, U., and Söderberg, G. (1982). Adjuvant interferon treatment of human osteosarcoma. In *Recent results in cancer research 80* (ed. G. Mathé, G. Bonadonna, and S. Salmon). pp. 103–7. Springer, Berlin.

Toy, J. L. (1983). The interferons. *Clin. Exp. Immunol.* **54,** 1–13.

Ueda, S., Hirakawa, K., Nakagawa, Y., Suzuki, K., and Kishida, T. (1983). Brain tumours. In *Interferon and cancer* (ed. K. Sikora) pp. 129–139, Plenum Press, New York.

Valter, S. (1977). The study of the effect of interferon on the radiosensitivity of cells. *Radiobiologia* **17,** 105–8.

Index